ENGINEERING DRAFTING and SOLID MODELING with SILVERSCREEN®

Y. C. Pao
Professor of Engineering Mechanics
University of Nebraska
Lincoln, Nebraska

In Consultation with
Michael Foltz
Director of Technical Support
Schroff Development Corporation
Mission, Kansas

CRC

CRC Press
Boca Raton Ann Arbor London Tokyo

Library of Congress Cataloging-in-Publication Data

Pao, Y. C.
Engineering drafting and solid modeling with silverscreen / by Y. -C. Pao, Michael E. Foltz.
p. cm.
Includes bibliographical references and index.
ISBN 0-8493-4471-9
1. Mechanical drawing. 2. Engineering models. 3. SilverScreen. I. Foltz, Michael E. II. Title.
T353.P193 1992
604.2—dc20 92-33595
CIP

Direct all inquiries to CRC Press, Inc., 2000 Corporate Blvd., N. W., Boca Raton, Florida, 33431.

International Standard Book Number 0-8493-4471-9

Library of Congress Card Number 92-33595
Printed in the United States 1 2 3 4 5 6 7 8 9 0
Printed on acid-free paper

Preface

Traditionally, technical drawings are in the form of blueprints. Computers have changed that form of presentation and we are now using the display monitors in abundance to give deteails of technical drawings in black and white, or, some may prefer in colors! Since computer graphics has been greatly advanced in the past decades to allow an object to be viewed from any selected angles, we are now often asked to present three-dimensional pictures of the object by **interactive control** of the viewing angle.

In the meantime, all real-world or man-made objects previously depicted using **wireframe models** are now being gradually replaced by using three-dimensional **solid models**. This text is designed for those who wish to learn about solid modeling. Leading towards the topics on solid modeling, the essential elements of engineering drafting and wireframe modeling are also discussed.

The solid modeling software **SilverScreen** is selected for the reason that it is inexpensive, versatile, and easy to apply. It is menu-driven and provides split-screen, multiple-window display for viewing any three-dimensional object. Its Boolean manupulations (union, subtraction, and others) of solids, and hidden-line removal capabilities are superior to other commercial software packages of comparable cost.

This text is suited to the college freshmen in an introductory course covering engineering drafting and simple computer-aided design. Since both the BASIC and C language are applied in this text in conjuction with the SilverScreen commands, this book can as well be adapted as a companion text for an introductory computer programming course.

Chapter 1 covers the loading and unloading procedures of SilverScreen and introduces some of its main features that are applicable for both the basic drafting and solid modeling needs. Chapter 2 commences the computer-aided drafting of two-dimensional objects, which can later be edited and modified using the SilverScreen commands introduced in Chapter 3. Procedures for adding dimensions and notes to a two-dimensional drawing are described in Chapter 4. The commands for producing a hardcopy of the screen display, and for saving the current drawing as a drawing file on disk and for later retrieval are presented in Chapter 5.

Chapter 6 begins the construction of three-dimensional solid models. Manipulations of solids such as combining them or subtracting one from the other, and other operations are discussed in Chapter 7. Methods for handling auxiliary and section views of a three-dimensional object, such as dimensioning on an oblique plane, are presented in Chapter 8. Grayscale shading and color rendering of three-dimensional objects to enhance the perspective views are the topics of Chapter 9. In Chapter 10, the SilverScreen commands are employed in the BASIC and C programs to show how existing programs already written in either BASIC or C language can be expanded by incorporating the features of SilverScreen.

For the readers who are anxious to have a fast track for a *short* but *complete* (in the sense of modeling an object, detailing it, producing a hardcopy and saving the drawing as a disk file) *lesson*, go directly to Sections 2.1, 2.4, 4.3, 5.2, 5.3, and 6.3.

This work is a part of my continuing involvement in the field of Computer-Aided-Engineering (CAE). It complements my other books *Introduction to Digital Computer Plotting* (coauthored by Professor Thomas C. Smith), *Elements of Computer-Aided Design and Manufacturing, CAD/CAM,* and *A First Course in Finite Element Analysis*. The present book covers engineering drafting and solid modeling. The former is indispensable in architectural and engineering design while the latter plays an important role not only in both the design and manufacturing but also in the preprocessing and postprocessing phases of finite-element analysis.

I participated in a short course on SilverScreen taught by Michael Foltz in the winter of 1990 and was very much impressed by the capabilities of this solid modeling software package designed by the Schroff Development Corporation. True, there are other more sophisticated versions of solid modeling software packages available in the marketplace, such as the ones provided by the PDA Engineering and by Intergraph. However, based on my over 25 years of teaching experience of CAE, SilverScreen is best adaptable for classroom uses and for self-study with a personal computer. The other software needs high-end computers and are for more sophisticated applications.

In preparation of this textbook, many of my close friends, students, and colleagues who had helped provide topics, examples, and homework problems for my other CAE books in the past, have again made invaluable suggestions and given me numerous oral and written materials. To them, I am forever grateful.

It would be impossible to have this textbook reach such a satisfactory, temporary halt without the constant advice and assistance from Michael Foltz. To him, I owe my deepest gratitude. Also, I would like to thank Stephen J. Schroff of the Schroff Development Corporation who has always been very helpful and patient when I so frequently called the SilverScreen office asking for clarification and assistance.

Professor Michael H. Pleck of the University of Illinois-Urbana, Professor Daniel L. Ryan of Clemson University, and Professor George Swancutt of the University of Missouri-Rolla who reviewed the first draft of this text, have provided a lot of very constructive comments. I would like to extend to them my sincere thanks for their input which has enabled me to bring before the readers a text of much better quality.

The editorial staff of CRC Press, Inc. are so friendly and efficient. I find it a great pleasure to have worked with them. Especially, I would like to thank Russ Hall, Rosi Larrondo, and Suzanne Lassandro, with whom I have frequent conversations on the phone and every time I have experienced a most cheerful dialogue. I also thank Marjorie Bisbee of the Engineering Research Centers Office at the University of Nebraska-Lincoln who has helped me convert the postscript files into color plots needed in this text.

The field of computer graphics is advancing so rapidly. This textbook must undergo continuous changes by incorporating the latest techniques and methodologies, and equally important the users' suggestions. So please continue to send in your critques and suggestions for betterment of this book.

Y. C. Pao

Contents

1 Introduction

1.1 ENGINEERING DRAFTING

It has been a long tradition that when one needs to make a machine part, a drawing such as Figure 1.1.1 must be prepared by a **draftsman**. On this drawing, all the details about the measurements of that part must be precisely specified so that it can be taken to the shop for a machinist to make it accordingly.

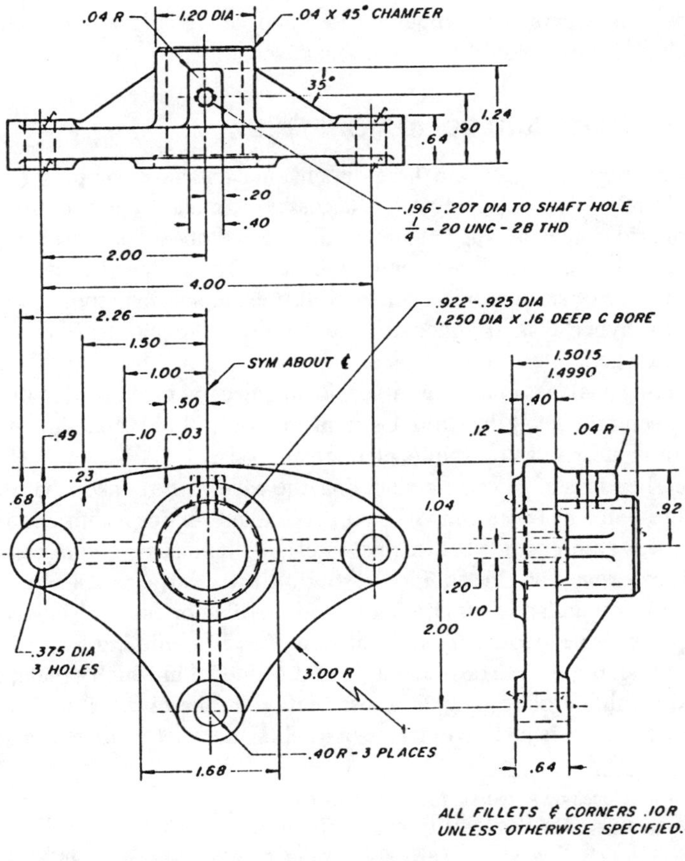

Figure 1.1.1. (Courtesy of General Motors Corporation)

Draftsmen still use pen or pencil to produce architectural, engineering, and business drawings on a drafting board. They are aided by the drafting instruments such as T-square or rulers with appropriate scales, french curve, or spline, protractor, divider, and compass. After a drawing has been finished, copies of the so-called **blueprint** are usually made for dissemination. But such practice is being increasingly replaced with a modern **computerized** method of using drafting software and producing a needed drawing on a display monitor by commands entered through keyboard and other peripheral devices such as mouse, lightpen, and graphic tablet.

Also, network arrangement is gradually eliminating the need of blueprint copies. The finished drawings are stored as files in computer memory and/or disk. With a work station in the shop, a remote terminal connected to a time-sharing computer enables the machinist to have access to the drawing files and to view on his/her display monitor the geometric specifications of the part to be made.

In this textbook, a Computer-Aided Drafting (**CAD**) software package called **SilverScreen** is applied for engineering drafting needs. There are other notable CAD packages such as **AutoCAD**[1] and **VersaCAD.**[2] SilverScreen is selected because compared to other packages it has better developed capabilities as far as solid modeling is concerned, it is personal-computer based and hence is lower in cost, and, most importantly, it is best suited for instructional uses.

1.2 WIREFRAME AND SOLID MODELING

Almost all real-world objects are three-dimensional and have complex geometry. For design and analysis of such objects, it is necessary to make some simplification about their geometric shapes by approximating them as composed of basic solids such as plates, cylinders, prisms, cones, and spheroids. These constituent solids can be generated by use of straight lines, spheres, ellipses, or special geometric curves such as parabolas, hyperbolas, trigonometric functions, and cycloids. For example, gear teeth are often generated with cycloids.

Engineering analyses of a three-dimensional object by investigating its constituent parts is the central idea of the **finite element method.**[3] Solid modeling of these parts is a topic intimately related to finite element analysis.

In the earlier days of geometric modeling, the **wireframe** concept has prevailed. A cube is represented by connecting its 8 corners, called **vertices**, with 12 straight lines. Evidently, it is ambiguous whether this wireframe represents a wire cage or a solid cube! Only in recent years, the addition of **shading** to the surfaces of a wireframe makes it possible to distinguish whether a framework is filled or empty. The availability of **color** graphics also expedites the introduction of **solid modeling**.

For example, Figure 1.2.1a shows a wireframe model in which a plate, a truncated prism, and a column of rectangular cross section are involved. It is difficult to tell whether the structure is just a wireframe or a solid, and whether there is a disk stacked

---

[1]**AutoCAD** is developed by Autodesk, Inc., Sausalito, CA.
[2]**VersaCAD** is developed by Versacad Corporation, Huntington Beach, CA.
[3]See, for example, Y. C. Pao, *A First Course in Finite Element Analysis*, Allyn & Bacon, Boston, 1986.

on the plate or there is a cutout. After shadings are added as shown in Figure 1.2.1b, it becomes a solid model and will not be mistaken for a skeleton of connected wires without fills.

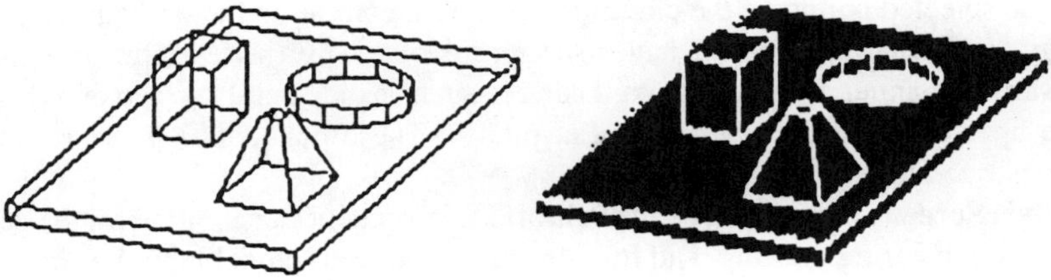

(a) (b)

Figure 1.2.1.

Sometimes, a wireframe model can be improved by removing the **hidden lines** as illustrated in Figure 1.2.2. Still, it is certainly not as good a presentation of the object as is shown in Figure 1.2.1b, which is produced by a solid-modeling approach.

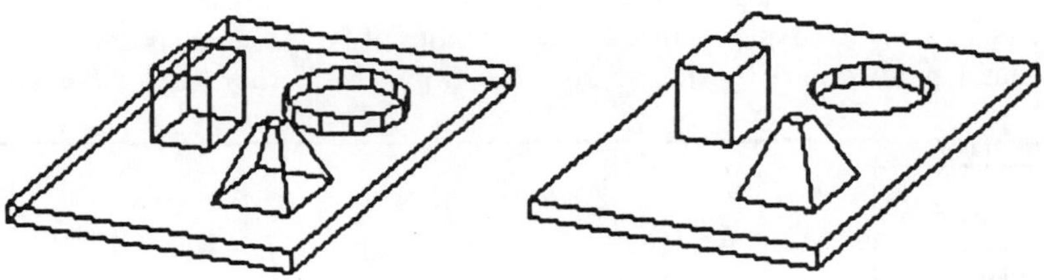

Figure 1.2.2.

In this textbook, we will explain in detail how solids can be generated by various techniques. For example, in Figure 1.2.3a two wireframes are generated by application of the software **SilverScreen**. It shows a rectangle and a circle displayed as a parallelogram and an ellipse when they are viewed perspectively. They can be connected to

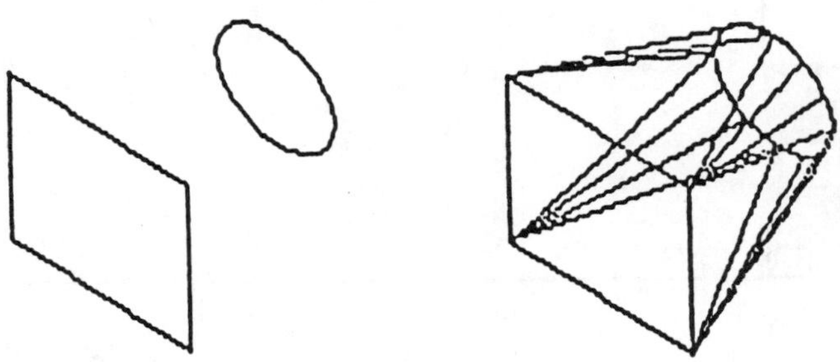

Figure 1.2.3.

generate a three-dimensional wireframe, shown in Figure 1.2.3b, by a **sweep** process of adding the lateral lines. To change it into a solid model, edge and surface colors can be added through interactive operations of SilverScreen.

We observe from Figure 1.2.3b that the four corners of the rectangle are connected to some selected points on the circumference of the circle. In fact, all curves such as the circle shown in Figure 1.2.3 have to be approximated by a series of linear segments. In a similar manner, a smooth curved surface such as the lateral surface of a cone has to be approximated by a series of flat planes. These concepts will be repeatedly demonstrated throughout this book.

SilverScreen is developed by the Schroff Development Corporation for the general need of engineering drafting, and for wireframe and solid modelings. We shall apply it throughout this book to demonstrate how three-dimensional objects can be represented using basic solids such as plates, spheres, cylinders, cones, and polyhedrons.

1.3 LOADING THE SILVERSCREEN SOFTWARE

SilverScreen can be loaded onto a computer by entering **Silver** from a keyboard to obtain an initial display (Figure 1.3.1) on screen. Subsequent actions are to be taken by entering the command characters from the keyboard or by pressing the control buttons on a mouse. Extensive applications of SilverScreen will be presented in the ensuing chapters. A thorough explanation of all capabilities of SilverScreen is documented in an official publication[4] by the Schroff Development Corporation. However, the

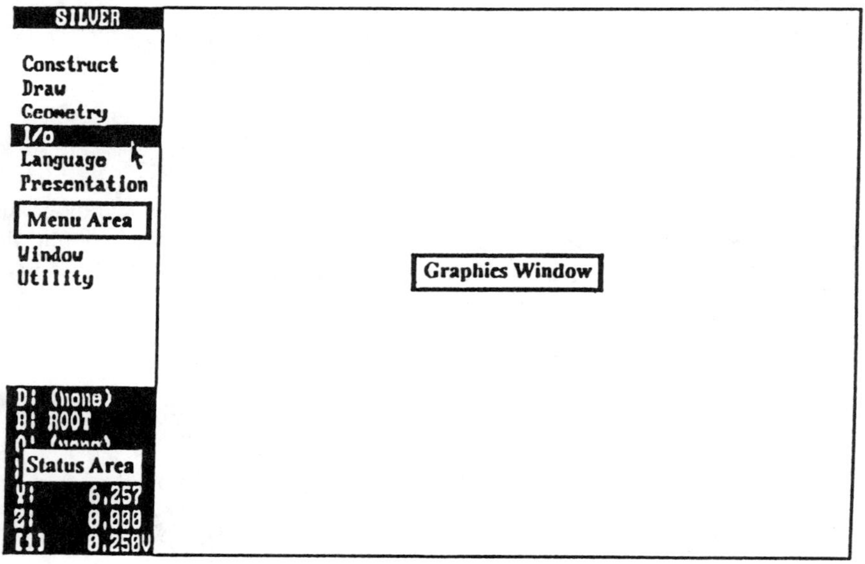

Figure 1.3.1.

[4] *SilverScreen™, Modeling Software*, Reference Manual Revision 1.0, Schroff Development Corporation, Shawnee-Mission, KS, 1989; Revision 2.1, 1990.

Reference Manual is more suited for experienced users rather than for beginners. This textbook is intended to gradually introduce the most often applied procedures available in **SilverScreen** for generations of lines, curves, wireframes, and solids by use of simple examples.

Here, as an introductory example, let us apply **SilverScreen** to draw an isosceles. First, we either enter a character **D** (for **Draw**) from the keyboard or move the mouse so that the blackened, rectangular bar presently located at **I/o** in Figure 1.3.1 is lifted to the location when the word **Draw** is displayed. This results in a change in the **Menu Area**, as illustrated in Figure 1.3.2. Next, we press the **Enter** key or the left button of the mouse to obtain a new menu as shown in Figure 1.3.2b. It indicates that there are many options available for **DRAW**.

Since our intention is to draw an isosceles, which is one type of polygon, we either enter a character **P** from the keyboard or maneuver the mouse so that the blackened rectangular cursor is positioned at the location where the word **Polygon** is displayed. Again, pressing the **Enter** key makes the screen to display a new menu as shown in Figure 1.3.2c. The size of a desired isosceles will then have to be specified by answering a series of prompted questions on the screen. Briefly, we describe the three steps as

Draw Polygon Isosceles

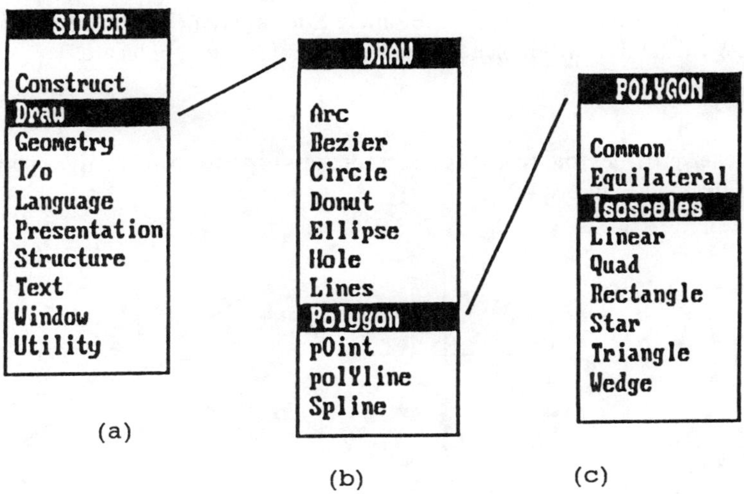

Figure 1.3.2.

Each command has a character indicated in uppercase and can be entered via the keyboard by pressing the uppercase character. For example, use **D**, **Y**, and **W** for **Draw**, **polYline**, and **Wedge**, respectively.

1.4 SOME PRELIMINARY NOTES ABOUT SILVERSCREEN

To apply the SilverScreen software, one may first want to specify what is the unit to be used in constructing an object. Some may prefer the British system of inches and feet, whereas others may like meters and centimeters better, or just generic terms. By consecutively requesting the menus of *Utility* and *uNits*, and then selecting the *Measure* option, the *Menu Area* will show the sequential prompts listed in Figure 1.4.1a, b, and c, respectively.

Utility	uNits	Measure
Clear	Denominator	Centimeters
Execute	Measure	Feet
Help	Notation	Generic
Lifestyle	Precision	Inches
Multi-window		Meters
Penstyle		miLlimeters
Starup		
Text-edit		
Undo		
uNits		
(a)	(b)	(c)

Figure 1.4.1.

One simply makes a selection from the (c) list by entering, for example, **G** for **Generic** from keyboard and pressing the **Enter** key. Hereon, all data entered will be in generic units until another unit change is made. In the **Status Area** (Figure 1.3.1), the rows showing *X*:, *Y*:, and *Z*: will indicate what the current selected units are.

CURSORS

SilverScreen provides four types of cursors for users to perform interactive operations. They are listed in Figure 1.4.2.

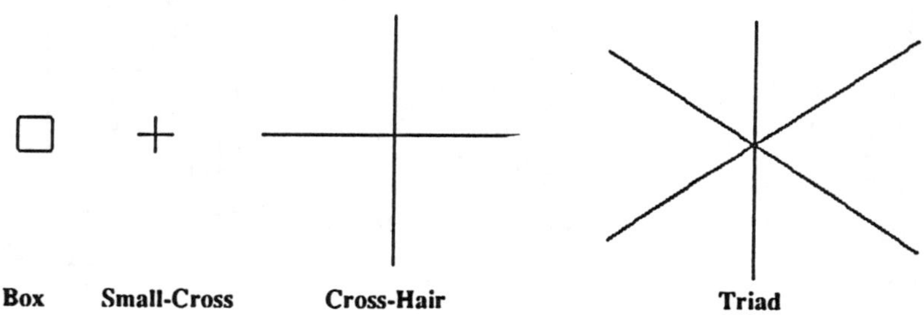

Figure 1.4.2.

Table 1.4.1 lists how the cursor can be moved by pressing various keys on the keyboard. It can also be moved by using a peripheral pointing device such as a mouse.

There are other keys that can be pressed to achieve different cursor movements, but those listed in Table 1.4.1 are used most often. The amount of the movement is

Table 1.4.1

Desired cursor motion	Key to be pressed
Move on positive *X*-axis	Right arrow
Move on negative *X*-axis	Left arrow
Move on positive *Y*-axis	Up arrow
Move on negative *Y*-axis	Down arrow
Move on positive *Z*-axis	Plus (+)
Move on negative *Z*-axis	Minus (-)

indicated in the last row of the **Status Area** and can be increased and decreased by pressing the **Ins** (Insert) and **Del** (Delete) keys, respectively.

The **box** and **small cross cursors** are used to mark where a point has been specified on screen. The **cross-hair cursor** is used in connection with a two-dimensional planar view while the **triad cursor** is used in connection with a three-dimensional **isometric view** of an object. Figure 1.4.3 shows a four-way, split-screen display of a block with five holes. Each screen has a set of cross-hair or triad cursor. For a color monitor, **SilverScreen** uses three different colors, cyan, yellow, and red, for the *X*-, *Y*-, and *Z*-axis, respectively.

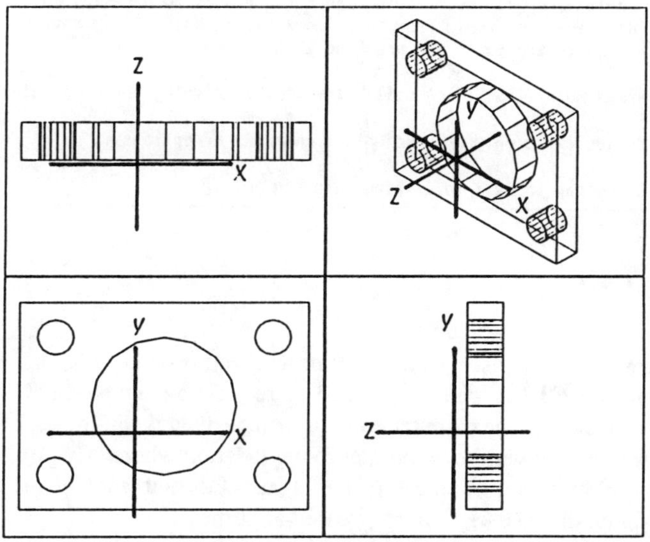

Figure 1.4.3.

User can select a preferred cursor by consecutively requesting for the **Construct** menu, the **Type** and **Cursor** commands, and then make a decision from the displayed list of Box, Cross-Hairs, Small Cross, and Triad (one of them is missing because it is the current cursor). Also, a quick change of cursor can be made by pressing the **ALT** (Alternate) and **C** keys simultaneously.

HELP

From the above brief introduction about various functions of **SilverScreen**, it is evident that the user needs to memorize how the available menus and commands should be applied, and which key should be pressed to request a certain task to be performed. To help the user, SilverScreen has provided a **Help** command in the **Utility** menu as indicated in Figure 1.4.1a. By entering the character **H** or **h** and pressing the **Enter** key, the screen displays general information about **SilverScreen**'s arrangement on how to get help! (Figure 1.4.4).

```
Related topics: QUICKKEYS  Mouse
ENTER Display related topic        F3 Help index      F4 Print help      F10 Exit

HELP  Help information

Help  information is available for every command.  When in  menu  mode,
this  information may be displayed by highlighting the command  name  on
the  menu and pressing F1.   Once this information has  been  displayed,
there are several ways to display related help:

    - A  related topic can be displayed by highlighting that  topic  and
      pressing ENTER.  (The space bar can be used to select the topic.)

    - An index of all help topics can be displayed by pressing F3.  Once
      displayed,  a topic from this index can be selected by moving  the
      arrow to the topic and pressing ENTER.

In command mode, the F1 key will automatically display the Help Index.

Help is available only when selecting commands or options.

The F2 key can be used to display Quickkey help.
```

Figure 1.4.4.

For example, if the user wants to know about various cursors available in **SilverScreen**, the **F3** key can be pressed to request an index list. On screen, all commands starting with the **Construct** menu will be displayed. The user then uses the Down Arrow key to lower the ⇒ pointer to the location where "Cursor" is listed and press the **Enter** key to request the explanation about the subject Cursor. To *exit* from the **Help** display, the **F10** key can be pressed anytime.

Alternatively, the user can ask for help anytime during the command mode by pressing the function key **F1** as indicated in Figure 1.4.4. Since they provide quick services, the function keys **F1** through **F10** in combination with the **Shift**, **Ctrl** (Control), and **Alt** (Alternate) are referred to as the **Quickkeys**. The assignments of these quickkeys for performing various **SilverScreen** tasks can also be displayed on screen by pressing the **F2** key as suggested in Figure 1.4.4. The resulting display is

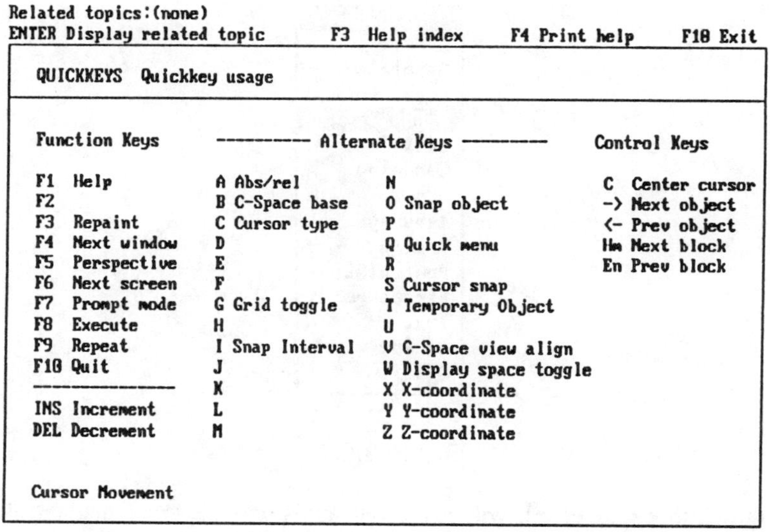

Figure 1.4.5.

shown in Figure 1.4.5. It shows that as previously mentioned, the **Alt** and **C** keys can be pressed to request a quick change of the cursor type and the **F1** key can be pressed to obtain a help display.

1.5 RECALLING SILVER MENU AND UNLOADING SILVERSCREEN

The **Silver** menu (Figure 1.3.2a) is the main menu of **SilverScreen**. The user can return to this menu by interrupting operation at anytime by pressing the **Esc** (**Escape**) key. To *terminate* the **SilverScreen** software, the **Esc** key can be pressed to return to the main (**Silver**) menu and then the **F10** key (10th Function Key) be pressed to obtain a prompting message on the first line of the screen: "Type **Q** to quit."

Only introductory comments about **SilverScreen** are given in this chapter to enable the interested readers to start application of the **SilverScreen** software for engineering drafting and solid modeling as quickly as possible. In the ensuing chapters, other **SilverScreen** commands will be introduced through presentation of a large collection of sample drawings, and examples of constructions of various solid models. It is believed that only by continual hands-on applications of the **SilverScreen** software can readers continue to have a keen interest in this subject.

SilverScreen is being continually refined. We will explain the major changes that have been made in the new version when it becomes available. Revision 2.0 has an **Annotate menu** (Figure 1.5.1) that provides improved methods for detailing (dimensioning and adding text lines to a drawing) of two- and three-dimensional objects.

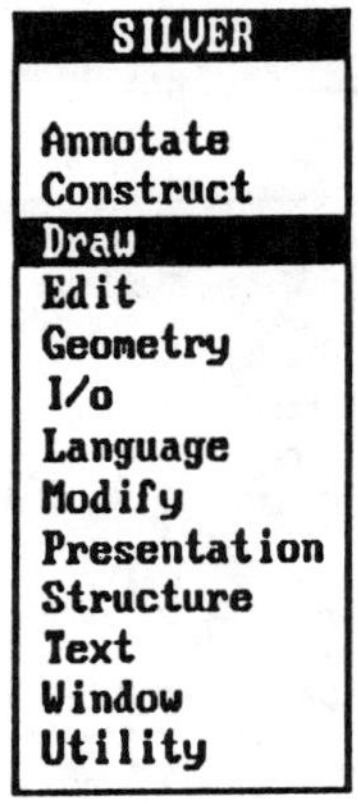

Figure 1.5.1.

In this textbook, all efforts will be made to apply only the fundamental techniques of **SilverScreen,** which are common in all versions of **SilverScreen**. In this way, users who possess an older version will then not be affected in their implementation of engineering drafting and solid modeling.

2 Two-Dimensional Drawings

2.1 INTRODUCTION

Anxious to start drawing pictures? To do so, one must know where the executable file **Silver.EXE** of the **SilverScreen** software package is resided. If **Silver.EXE** is stored on a hard drive C, we simply enter the request

C > **Silver**

By selecting the **Draw** option in the main menu, **Silver** shown in Figure 2.1.1a, of **SilverScreen**, we are ready to draw! Figure 2.1.1b shows that a variety of basic geometric shapes such as lines, circles, and polygons called **primitives** is available for users to employ to generate any combined geometric shapes. In this chapter, we first describe the generation of geometric curves on a plane involving points that can be specified with two coordinates, (X,Y). Spatial curves involving points which need to be defined with three coordinates, (X,Y,Z), will be discussed in the ensuing chapters.

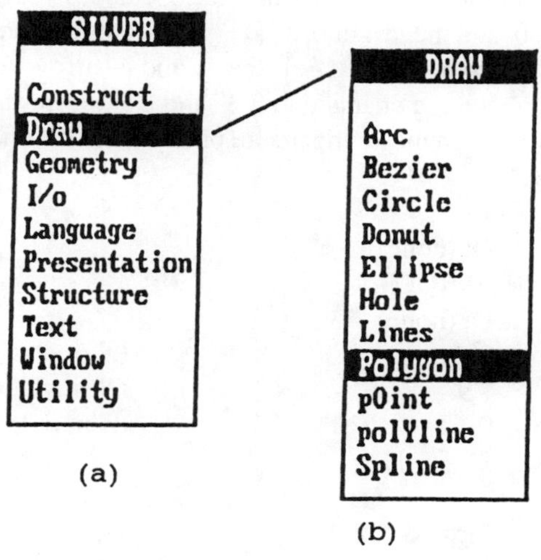

Figure 2.1.1.

The fixed coordinate system, called World-space (**W-space**), will be used in this chapter; other systems, C-space and E-space, which are also provided by **SilverScreen,** will be introduced later. Examples of drawings using the points specified by both **Absolute** and **Relative** coordinates will be presented.

2.2 SINGLE LINES

Let us start by drawing a line on the *X–Y* plane. A line can be drawn by specifying either two points or one point and a slope for that line. The following sequence of **SilverScreen** commands enables a **Single Line** to be drawn:

Draw Lines Single

It results in a prompting message at top of the screen

Mark or enter start of first line:_

The user has **three options** to specify the coordinates of the starting point of a desired line: (1) by typing the *X* and *Y* coordinates, (2) by moving the pointing device (mouse or light pen) to shift the + **cursor** on screen so that the intercept of the horizontal and vertical lines of the cursor is at the desired location, or (3) by pressing the rightward, leftward, upward, and downward **arrow** keys on the keyboard to increase and decrease the *X* value, and increase and decrease the *Y* value, respectively. When the second and third options are selected, the user will witness the changes of the *X* and *Y* values displayed in the **Status Area** (see Figure 1.3.1) located at the left lower corner of the screen. For example, Figure 2.2.1 shows that the + cursor is currently positioned at *X*=9′, Y=7′9″, and Z=0′ and the last line, 3″ [1]V, indicates that among other explanations, the increment is currently set at 3″. If the → and ← arrow keys on the keyboard are pressed, the *X* value will be changed to 9′3″ and 8′9″, respectively. If the upward and downward arrow keys on the keyboard are pressed, the *Y* value will be changed to 8′ and 7′6″, respectively.

D: (none)	
B: ROOT	
O: (none)	
9′	(X in blue)
7′9″	(Y in yellow)
0′	(Z in red)
3″ [1]V	

Figure 2.2.1

After the user has decided on a location (X_1,Y_1) for the starting point of a line, pressing the **Enter** key will result in a colored dot marked on the screen. In the meantime, the screen prompts with another message

Mark or enter end point:_
Length: *Dx*: *Dy*: *Dz*:

Again, any one of the above-mentioned three options can be selected to specify the ending point of the line. If arrow keys or mouse are used, the changes in *X*, *Y*, and *Z* values and the length of the line calculated from using the already specified beginning point and the present cursor intercept are displayed after *Dx*:, *Dy*:, *Dz*:, and Length:, respectively.

After the user has decided on a location (X_2,Y_2) for the ending point of a line, pressing the **Enter** key results in a line being displayed from the colored dot to the intercept of the + cursor. This is done in a **vector-graphics** way by lighting up all pixels (picture elements) along the line connecting the points (X_1,Y_1) and (X_2,Y_2). The equation used for displaying this line is

$$Y = Y_1 + (X-X_1)(Y_2-Y_1)/(X_2-X_1)$$

If the user finds the displayed line unsatisfactory, as prompted in the **Status Area,** the **F1** key can be pressed to *undo* the drawing. The line can be interactively specified again.

As an example, if a line is to be drawn from (3,4) to (5,6) on the *X–Y* plane (*Z*=0), we may respond via keyboard as follows:

Mark or enter start of first line: 3,4
Mark or enter end point: 5,6

The values entered are *unformatted* and for the *X* and *Y* coordinates, respectively, while the unmentioned third coordinate automatically imply that Z=0. Also, these values are *absolute* values in reference to the origin of the W-space.

The line can also be drawn using *relative* values. After the starting point (3,4,0) has been entered, the **cross-hair cursor** will be located at that point. **Alt** and **A** keys on the keyboard can be simultaneously pressed to specify that the ending point of the desired line is to be entered with coordinate values that are measured from the location of the current cursor. A message will be shown on screen

Coordinate entry mode is now **RELATIVE.**

To draw the same line as before, the ending point should now be entered as

Mark or enter end point: 2,2

Alt and **A** keys can be simultaneously pressed to change back to the **ABSOLUTE** mode.

The resulting display is shown in Figure 2.2.2. It is difficult to tell whether the line has been correctly drawn without some reference axes and scales. The **Grid** command in the **Construct** menu of **SilverScreen** can be effectively utilized here as illustrated in Figure 2.2.3 where a 41 × 41 matrix of dots is added to the drawn line. The origin is at the left, lower corner of the rectangular border. The horizontal and vertical increments for the dots are both equal to 0.25. To request this rectangular grid to appear on the screen, the commands to be entered are

Construct Grid Enable

The grid can be removed by the commands **Construct**, **Grid**, and **Disable**, or by pressing the **Alt** and **G** keys simultaneously.

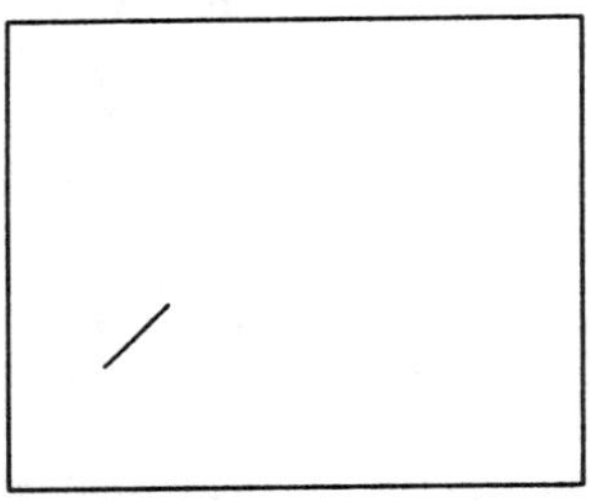

Figure 2.2.2.

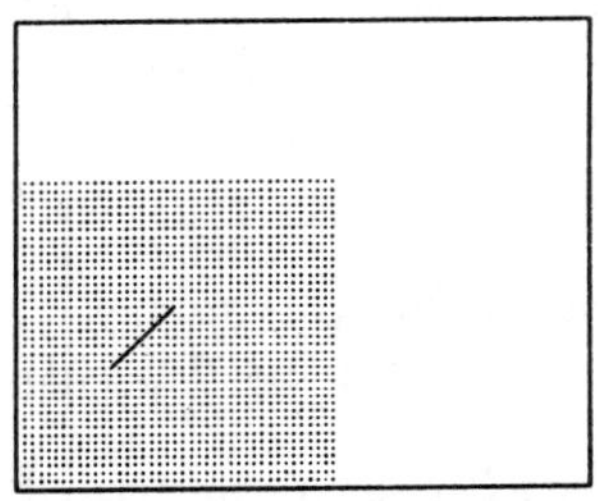

Figure 2.2.3.

Formatted coordinate values may also be used to specify the locations of the starting and ending points of a desired line. To draw the line shown in Figure 2.2.2, we may enter

Mark or enter start of first line: x3y4
Mark or enter end point: y6x5

or

Mark or enter start of first line: y4x3
Mark or enter end point: x2y2r

Notice that the latter set uses relative values and that in using formatted input, either order of x3y4 or y4x3 is acceptable.

2.3 CONNECTED MULTIPLE LINES

Multiple lines, such as the contour ABCDEFA shown in Figure 2.3.1 involving points all on the $z = 0$ plane can also be drawn using **SilverScreen** as follows:

Draw Lines Multiple
Mark or enter start of first line: 0.1875,0 (Point A)
Mark or enter endpoint: 0.8175,0 (Point B)
Mark or enter endpoint: 0.8175,0.25 (Point C)
Mark or enter endpoint: 0.5,0.5 (Point D)
Mark or enter endpoint: 0.5,1.25 (Point E)
Mark or enter endpoint: 0.1875,1.25 (Point F)
Mark or enter endpoint: 0.1875,0 (Point A)
Press F10 key to finish

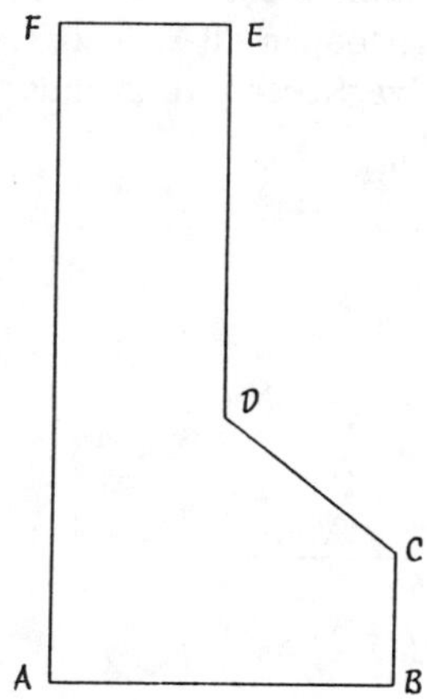

Figure 2.3.1.

As a second example, let us draw a star by entering the commands:

Draw Lines Multiple
Mark or enter start of first line: 2′,6′,0′
Mark or enter end point: 5′,6′,0′
Mark or enter end point: 5′,6′,0′
Mark or enter end point: 8′,0′,0′
Mark or enter end point: 7′,3′,0′
Mark or enter end point: 3′,3′,0′
Mark or enter end point: 5′,6′,0′
Press F10 key to finish

The resulting display on screen is shown in Figure 2.3.2.

A stairway may also be modeled as shown in Figure 2.3.3. The command entered are (Z coordinates all equal to 0′ if not entered)

Draw Lines Multiple
Mark or enter start of first line: 1′,1′
Mark or enter end point: 1′,9′
Mark or enter end point: 2′,7′
Mark or enter end point: 3′,8′
. . . . (repeatedly increasing *X* and *Y* coordinates by 1′)
Mark or enter end point: 9′,1′
Mark or enter end point: 1′,1′
Press F10 key to finish

For such a repetitious specification of coordinates of the points required in drawing a series of connected lines, using a loop in a **BASIC** or **C** program will be more efficient. Chapter 10 will address how **SilverScreen** commands can be incorporated into **BASIC** or **C** programs.

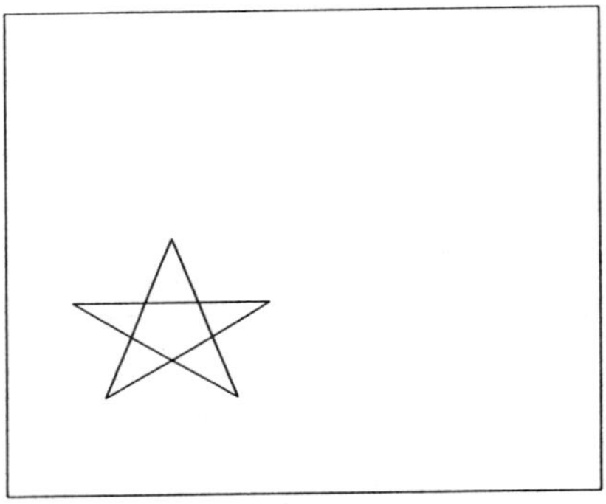

Figure 2.3.2.

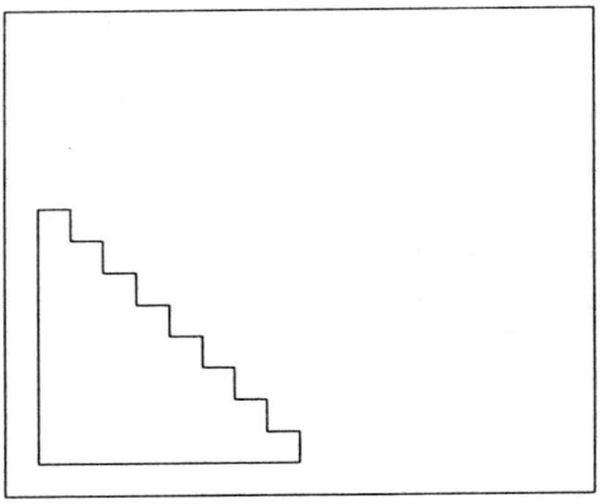

Figure 2.3.3.

2.4 POLYGONS

Since rectangles are the most commonly encountered two-dimensional geometric shapes, they can certainly be drawn using the multiple-line approach explained in the preceding section. But a speedier way of drawing it is by using the commands

Draw Polygon Rectangle

This allows the user to draw a rectangle by specifying the locations of two of its opposite corners. For example, to draw the rectangle in Figure 2.4.1 we may respond as follows:

Mark or enter corner of rectangle: 0,0
Mark or enter second corner of rectangle: 8,6

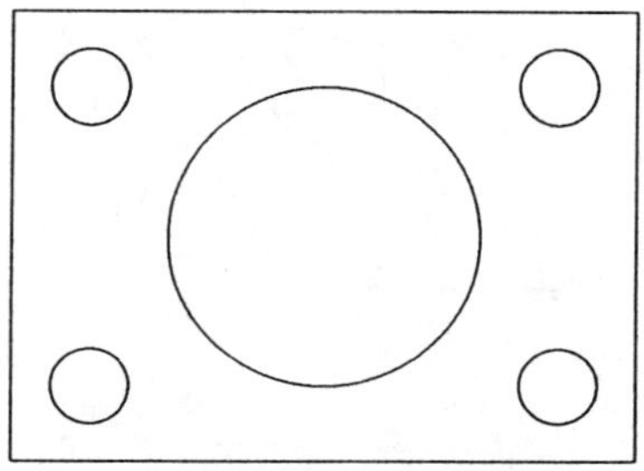

Figure 2.4.1.

The options for drawing polygons available in **SilverScreen** are

Common	**Equilateral**	**Isosceles**	**Linear**	**Quad**
Rectangle	**Star**	**Triangle**	**Wedge**	

As a next example of application, let us draw the equilateral polygons shown in Figure 2.4.2. To drawing a four-sided equilateral polygon, the commands are as follows:

Draw Polygon Equilateral
Enter number of sides in polygon: 4
Vertex (selected from **Edge** or **Vertex**)
Mark or enter center of polygon: 8′9″,11′6″,0
Mark or enter point for polygon: 8′9″,13′6″,0

A similar procedure can be tried to obtain the other polygons with increasing numbers of sides, (from 3 to 11) shown in Figure 2.4.2. Readers should also attempt to draw a polygon using the **Common**, **Star**, **Wedge**, and other options to gain hands-on experience.

Multiple rectangular shapes can be applied for modeling of various three-dimensional objects such as the window shown in Figure 2.4.3. Rectangles of three different sizes (4′8″ × 4′6″, 2′3″ × 4′2″, and 10 5/8″ × 1′3″) can be repeatedly drawn at properly selected locations using the commands **Polygons** and **Rectangle**. Since there are repetitions, we may want to draw only one and then copy and place it at other desired locations. This need of copying will be addressed in Section 2.6.

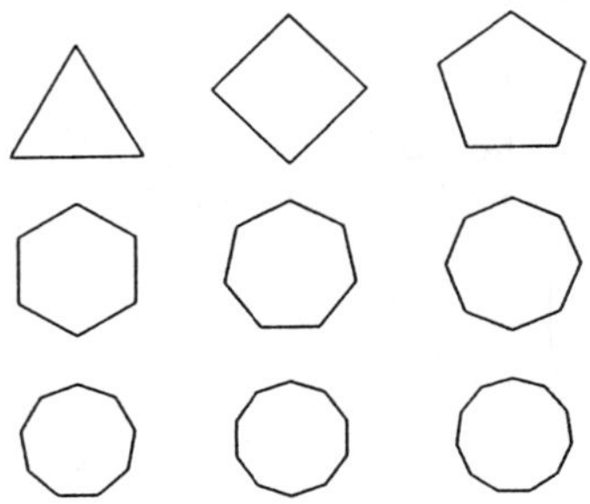

Figure 2.4.2.

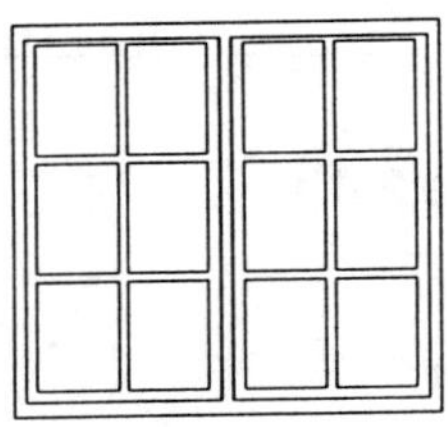

Figure 2.4.3.

2.5 CIRCLES

Having a rectangle already drawn in Figure 2.4.1, let us add a circle at the left lower corner by use of the commands

Draw Circle Radius

and by answering the subsequently prompted questions as follows:

Mark or enter center of circle: 1,1
Mark or enter radius point: 1.5,1

Notice that the radius point is a point on the circumference of the desired circle. If we denote the entered coordinates of the center as (X_c,Y_c) and those of the radius point as (X_r,Y_r), the circle is drawn using the equation

$$(X - X_c)^2 + (Y - Y_c)^2 = R^2 \tag{1}$$

In (1), R is the radius of the circle and can be calculated by using the formula

$$R = \left[(X_r - X_c)^2 + (Y_r - Y_c)^2\right]^{\frac{1}{2}} \tag{2}$$

When circles are requested, the **Circle** menu has options of drawing a circle by **Diameter**, **Radius**, **Points**, and **Tangent**. The above example shows the application of the **Radius** option. Let us draw the large circle in Figure 2.4.1 by the **Diameter** option. The command are

Draw Circle Diameter

and the circle is drawn with the responses of

Mark or enter start of diameter: 4,1
Mark or enter second point of diameter: 4,5

Notice that the two points specified are at the opposite ends of a diameter of the desired circle. If the coordinates of these two points are denoted as (X_1,Y_1) and (X_2,Y_2), the circle is then drawn with the equation

$$(X - X_c)^2 + (Y - Y_c)^2 = D^2 / 4 \tag{3}$$

In (3), D is the diameter of the circle and can be calculated by using the formula

$$D = \left[(X_2 - X_1)^2 + (Y_2 - Y_1)^2\right]^{\frac{1}{2}} \tag{4}$$

The coordinates of the center of the circle, (X_c,Y_c), needed in (3) are calculated by using the formulas

$$X_c = (X_2 + X_1)/2 \qquad \text{and} \qquad Y_c = (Y_2 + Y_1)/2 \tag{5,6}$$

REFINEMENT OF CURVES

Figure 2.5.1 shows that the circle is drawn by connecting a series of linear segments. If there are insufficient number of linear segments, the circle will appear like a polygon. Refinement can be made by selecting the **Display** command in the **Presentation** menu

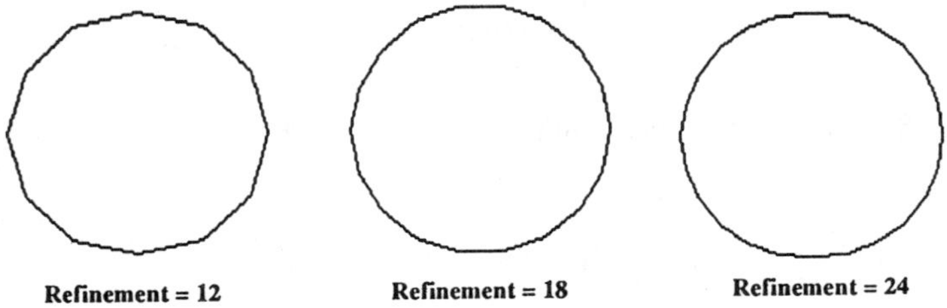

Figure 2.5.1.

(Figure 2.5.2) as follows:

Presentation Display Refine Circle screeN

Enter new refinement for (option):18

Figure 2.5.2.

If the current refinement arrangement is for using 12 linear segments, the above commands will result in a change from 12 to 18 linear segments. This refinement will remain effective until another refinement change is made.

2.6 COPYING A DRAWN PRIMITIVE

There are three other small circles in Figure 2.4.1 located at the lower right, and upper left and right corners that need to be drawn. Of course, we may draw them one by one using either the **Radius**, **Diameter**, **Points**, or **Tangent** option. But,it is an appropriate time to introduce the technique of using the **Copy** command in the **Geometry** menu

(Figure 2.6.1) of **SilverScreen**, which enables a drawn geometric shape to be copied and then placed at other locations.

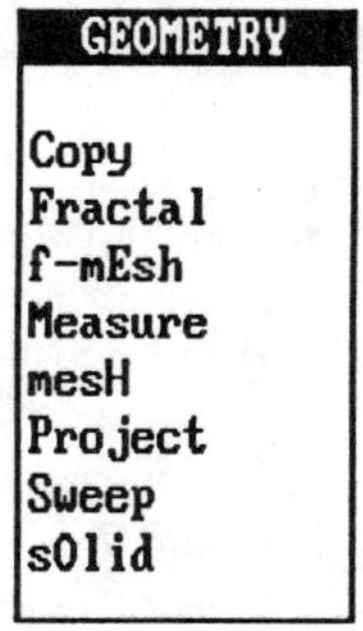

Figure 2.6.1.

To copy the small circle in Figure 2.4.1, we enter commands

Geometry Copy Primitive Cirlipse

which request the geometry of a circle or ellipse (primitives) to be copied. For single copying, such as copying the small circle at the left lower corner and placing it at the right lower corner, we move the + cursor in close vincinity of the small circle and then press the **Enter** key. This decides that between the two circles, one large and one small appearing on the screen, the small one is to be selected. Next, we select command **Single** and respond as follows:

Mark or enter base point for displacement
Enter base point: 1,1
(This selects the center of the circle as a base point.)
Enter second point: 7,1
(To place the copied circle, the base point is moved to a new location whose coordinates are $X = 7'$ and $Y = 1'$ and serves as the center of the circle.)

This single-copying procedure is to be repeatedly applied for drawing the two other small circles at the upper left and right corners in Figure 2.4.1. A better method for copying these four circles is to employ the **Matrix** copying method because the circles appear in matrix form of 2 rows and 2 columns. To do so, we enter commands

Geometry Copy Primitives Cirlipse Matrix

The needed responses are

Enter number of rows: 2
Mark or enter base point for displacement between rows: 1,1
(The center of the small circle to be copied is selected as a base point.)
Mark or enter a second point: 1,5
(The center of the copied circle is to be placed at the next row and at the location $X = 1'$ and $Y = 5'$.)
Enter number of columns: 2
Mark or enter base point for displacement between column: 1,1
Mark or enter a second point: 7,1
(The center of the copied circle is to be placed at the next column and at the location $X = 7'$ and $Y = 1'$.)

The smallest rectangle shown in Figure 2.4.3 is drawn repeatedly 12 times but placed at different locations. We can achieve the result by drawing one at the lowest, left corner of the left window. By copying it using a matrix of three rows and two

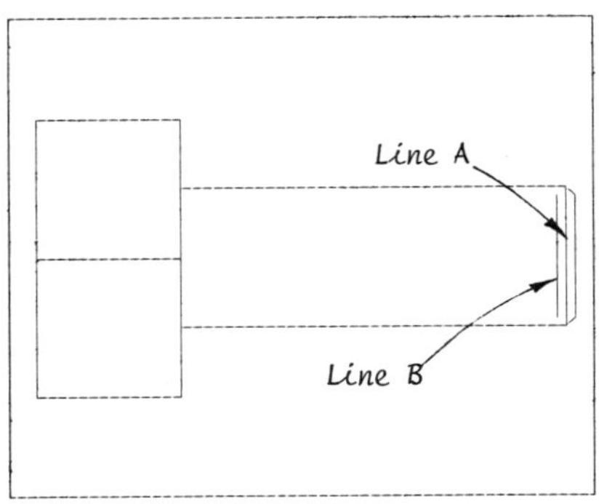

Figure 2.6.2.

columns, we complete the drawing of the left window. Next, a single copy of this left window can be made to produce the right window.

As another example of applying the **Copy** command of **SilverScreen**, we represent the *crest* and *root* of the threads in a standard hex bolt (Appendix B) by lines A and B, respectively, in Figure 2.6.2. First, to draw the other crest lines, the commands **Copy** and **Linear** can be effectively applied to produce a display as in Figure 2.6.3. A similar procedure for copying the root lines can be followed resulting in a final drawing as shown in Figure 2.6.4 for solid modeling of a bolt.

It is not intended to be exhaustive in application of **SilverScreen** for drawing two-dimensional pictures in this chapter. The materials introduced are merely enough for us to move on to the discussion of three-dimensional drawings, eventually leading toward the central topic of solid modeling. **SilverScreen** reference manual should be consulted if the user is interested in exploring the other drawing options.

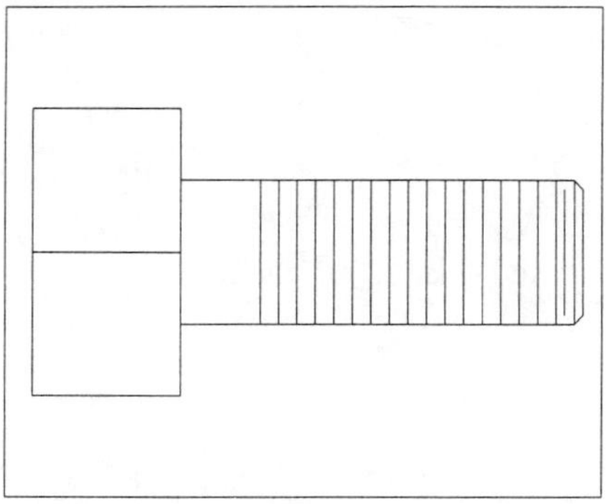

Figure 2.6.3.

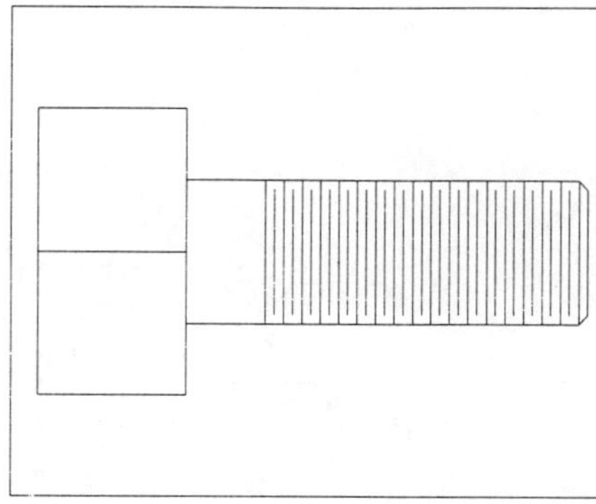

Figure 2.6.4.

2.7 DRAWING SMOOTH CURVED LINES AND CONTOURS

Frequently, we may have to draw not a complete circle but a circular arc. Or we may have to draw a smooth curve passing a set of specified points. The **Arc**, **Bezier**, and **Spline** commands in the **Draw** menu (Figure 2.7.1) of **SilverScreen** can be readily applied for such needs.

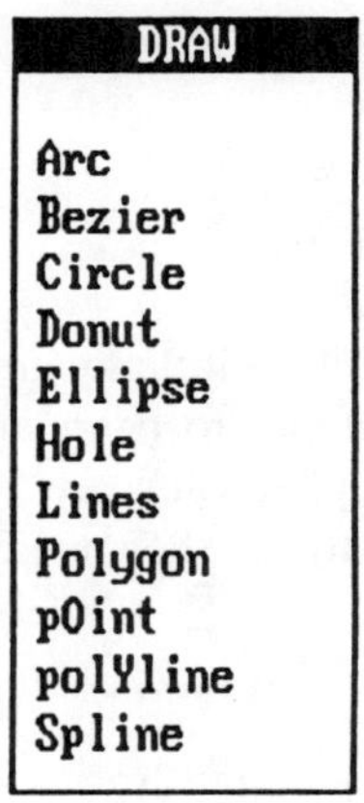

Figure 2.7.1.

ARC COMMAND

Suppose that we want to draw a semicircle. As shown in Figure 2.7.2, there are a number of ways that the **ARC** command in the **Draw** menu of **SilverScreen** can be applied. If one enters the commends **Draw** and **ARC**, the **menu area** will show four options: **Angles**, **Chord**, **Points**, and **Tangent**. The **Points** option is easiest to

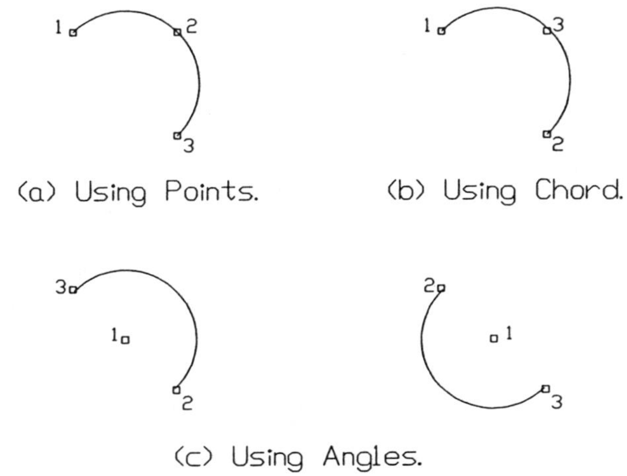

Figure 2.7.2.

understand. We simply specify three points that the desired circular arc should be passing through. To draw the arc shown in Figure 2.7.2a, we enter the commands and responses

Draw Arc Points
Mark or enter the first point: 3′,12′
Mark or enter the second point: 5′,12′
Mark or enter the third point: 5′,10′

It should be noted that actually the three little squares and the labels are not displayed on screen, they are added to depict the procedure and locations involved.

The **Chord** option should be applied with care. To obtain the same semicircle, the order of the three point is changed to that shown in Figure 2.7.2b. The commands and responses to be entered are

Draw Arc Chord
Mark or enter the first point of chord: 3′,12′
Mark or enter the second point of chord: 5′,10′
Mark midpoint of arc
Chord to midpoint: Angle: Radius:

In response to the last two lines prompted on top of the screen, arrow keys or the pointing device are to be used to change the location of the screen cursor to the desired location. Wherever the screen cursor is moved to, the perpendicular distance measured from the midpoint to the chord, angle, and radius is immediately calculated and interactively displayed after the three : signs. Pressing the **Enter** key finalizes the location of the midpoint. In the case of Figure 2.7.2b, the screen cursor should be moved to the location (5′,12′). If the midpoint is not placed, the resulting arc may not be a semicircle!

The **Angles** option should also be applied with care if the desired arc is a semicircle. Even the results are semicircles of the right size with the same diameter as shown in Figure 2.7.2(c), only the left semicircle is the desired one. The commands and responses to be entered are

Draw Arc Angles
Mark or enter center: 4′,9′
Enter starting angle: 5′,10′
Enter stop angle: 3′,12′

In fact, any point along the starting and stop angles can be specified to obtain the same result. However, the order of entering these two angles affects the outcome, as demonstrated in Figure 2.7.2c.

The **Tangent** option requires that some lines and arcs or circles are already drawn. The needed arc will be drawn tangent to two or three specified primitives. It is referred to as **Two-way** and **three-way** drawing of tangential arcs. A number of examples are shown in Figure 2.7.3.

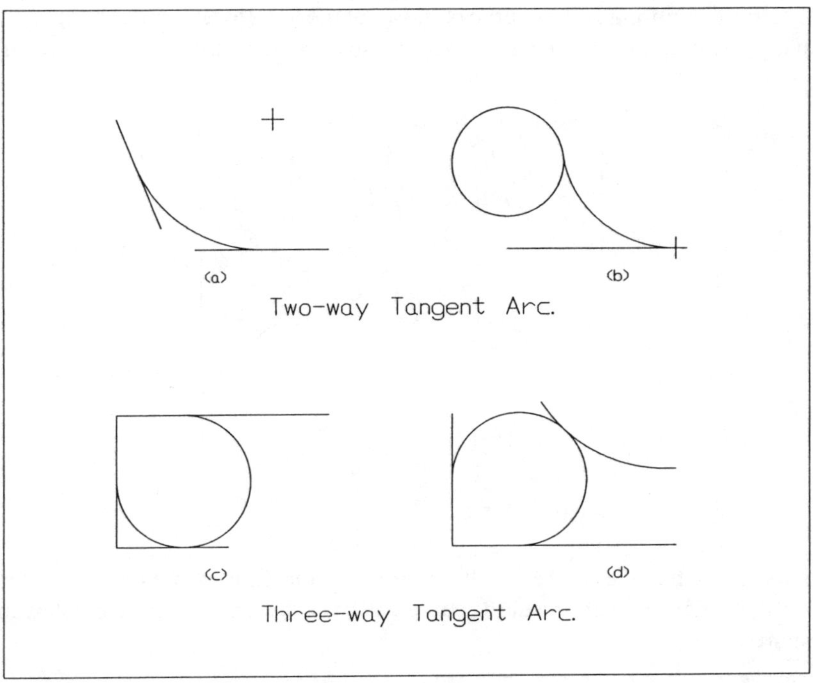

Figure 2.7.3.

BEZIER COMMAND

A smooth curved line can be drawn if a sufficient number of points on it are known. How accurate this curve is will depend on the number of points. **Bezier curves**[1] can be employed to fit the specified points smoothly. **SilverScreen** provides such a need in drawing curved lines and contours. A **Bezier** command is made available in the **Draw** menu for this purpose. There are two options in drawing Bezier curves: **Open** and **Closed**. Consider the case of knowing that four points are on a smooth curved line. We may draw a smooth curve passing these four points by entering the commands and responses

Draw Bezier Open Single
Mark or enter point: (enter the coordinates of the first point)
Mark or enter point: (enter the coordinates of the second point)
Mark or enter point: (enter the coordinates of the third point)
Mark or enter point: (enter the coordinates of the fourth point)
Press **F10** key to end.

A sample, resulting Bezier curve is shown at the left, lower corner in Figure 2.7.4. Actually, the four dots on the curve are not displayed; they are shown to help indicate

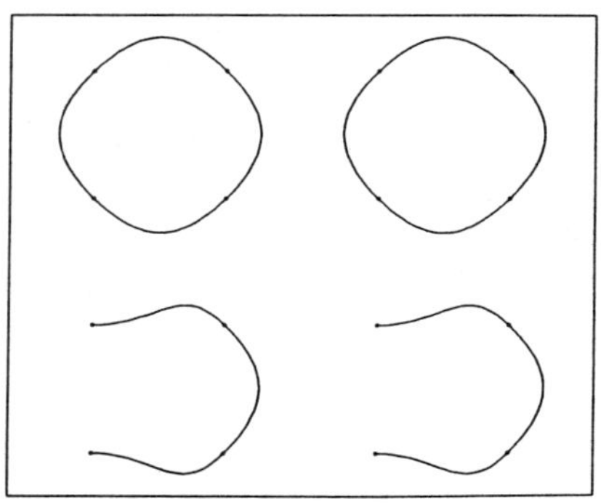

Figure 2.7.4.

their locations relative to the resulting curve. If the **Closed** option is selected for drawing a Bezier curve, the resulting curve will be as shown at the the upper, left corner in Figure 2.7.4.

---

[1]See, for example, Gerald Farin, *Curves and Surfaces for Computer Aided Geometric Design, A Practical Guide*, Academic Press, San Diego, CA., 1988.

Two control options, **Single** and **Double**, are available for drawing either open or closed Bezier curves with single and double control. The curves shown in the left and right columns in Figure 2.7.4 are the results of selecting single and double control, respectively. Here, there seems to be no difference in the resulting Bezier curves in selecting single or double control, but in Chapter 3, we will show how Bezier curves can be reshaped and the single and double controls play an important role in the reshaping process.

SPLINE COMMAND

Another way of fitting a number of specified points by a smooth curve is by applying the spline equations.[2] Again, both **Open** and **Closed** options are availbale in **SilverScreen**. The commands and responses to be entered are

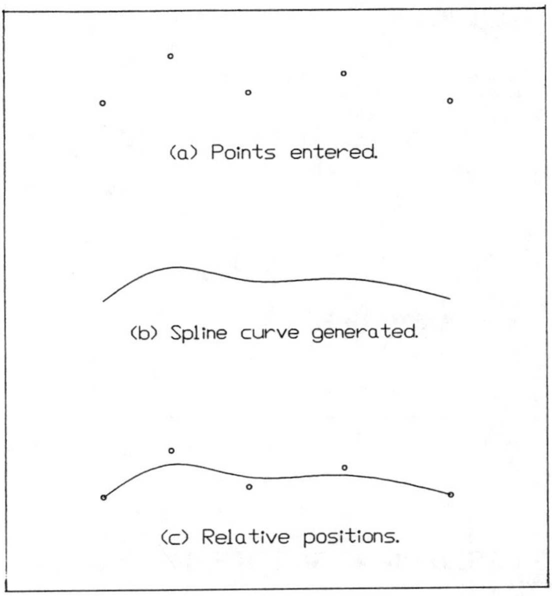

Figure 2.7.5.

Draw Spline Open
Mark or enter point: (enter the coordinates of the first point)
Mark or enter point: (enter the coordinates of the second point)

---

[2]See, for example, P. Lancaster and K. Salkauskas, *Curve and Surface Fitting, An Introduction*, Academic Press, San Diego, CA., 1986.

.

(repeat entering coordinates of other points)

.

Press **F10** key to end.

A sample, resulting open-spline curve is shown in Figure 2.7.5. If the **Closed** option is selected, the resulting spline curve is presented in Figure 2.7.6.

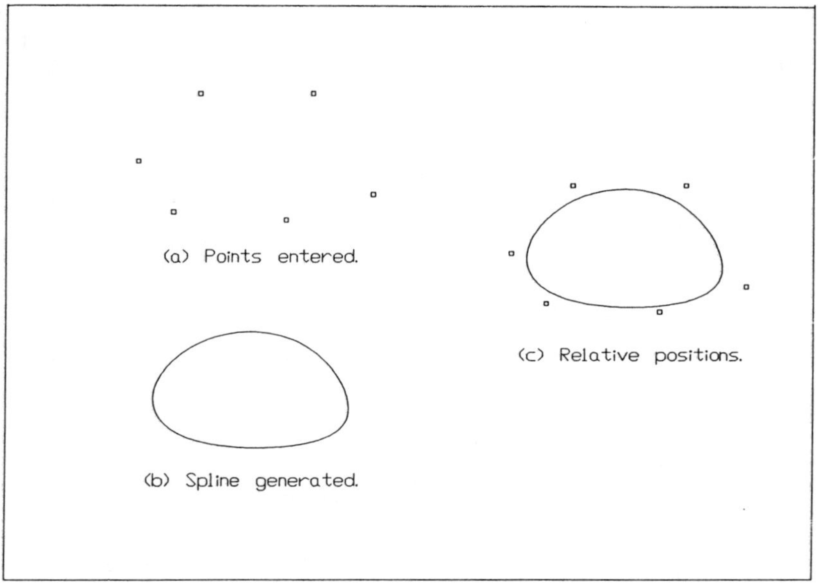

Figure 2.7.6.

2.8 DRAWING BASED ON A USER-DEFINED FUNCTION, F-MESH APPLICATIONS

The user may specify a series of points to be connected to display a desired graph and often the required points can be generated by use of a mathematical equation. This means that a program is needed to define that equation and then uses it to generate a desired number of points. **SilverScreen** recognizes this need and has a command called **F-Mesh** available that first reads an **ASCII** (American Standard Codes for Information Interchange) file and then displays the user-defined graph.

For example, if we want to generate a 10 by 10 square mesh, the following program can be written and stored in the subdirectory **EX** of **SilverScreen**. This file must have an extension **.PM** to make it accessible to the **F-Mesh** command.

'F-Mesh Program SquareMh.PM—A Square Mesh.
XV: 1 to 10 by 1
YV: 1 to 10 by 1
$X = XV$
$Y = YV$
$Z = 0$

The above program shows that an ' (apostrophe or single quote) can be used for the commenting purpose. The variables *XV* and *YV*, very much like **DO** loops in **FORTRAN** or **FOR-NEXT** in **C** and **BASIC**, serve as **parameters**. These two parameters are both to vary from 1 to 10 with a constant increment of 1. *X*, *Y*, and *Z* are the coordinates of the grid points of the square mesh. *X* and *Y* are to vary with the parameters *XV* and *YV*, respectively. The $Z = 0$ statement restricts the square mesh to be planar.

The process for applying the **F-Mesh** command to display square mesh shown in Figure 2.8.1 goes as follows:

Geometry **f-mEsh**
Enter name of file containing functional definition:
SquareMh
Select or enter horizontal color: 15
Select or enter vertical color: 15

Next, we include two more examples of **F-Mesh** application. Program **PolarMh.PM** generates a mesh in polar coordinates and the resulting plot is shown in Figure 2.8.2.

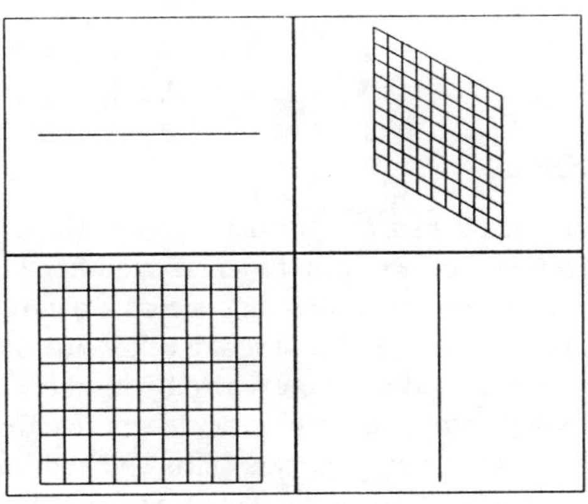

Figure 2.8.1.

'F-Mesh Program PolarMh.PM — a polar mesh.
RV: 3 to 8 by 1
CV: 0 to PI/2 by PI/18

$X = RV*\cos(CV)$
$Y = RV*\sin(CV)$

The other program is **ParabMh.PM,** which generates a mesh using three parallel parabolas. The display is shown in Figure 2.8.3.

```
'F-Mesh Program ParabMh.PM — draws parabolic mesh.
a2 = 1
a3 = -.04
t: 1 to 11 by 1
a1: 0 to 4 by 2
x = t
y = a1+a2*t+a3*t^2
```

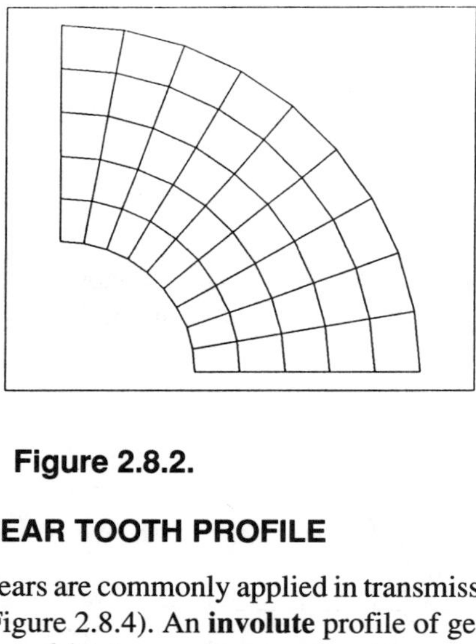

Figure 2.8.2.

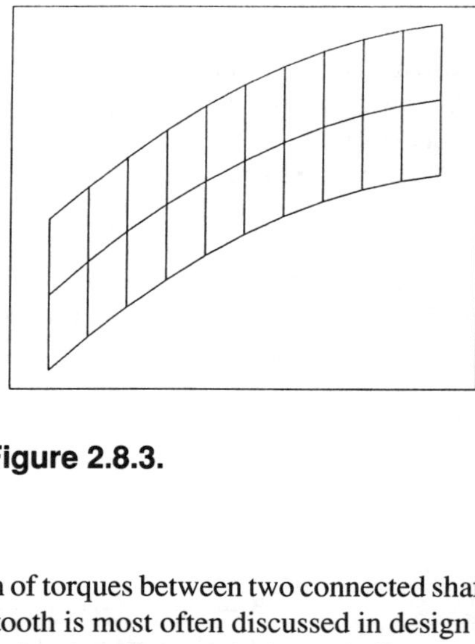

Figure 2.8.3.

GEAR TOOTH PROFILE

Gears are commonly applied in transmission of torques between two connected shafts (Figure 2.8.4). An **involute** profile of gear tooth is most often discussed in design of gears.[3,4] It is a profile developed by unwrapping a cord around a base circle.

Very briefly, we include below a **.PM** program called **Involute** that draws an involute profile of a spur gear and can be executed with the **F-Mesh** command of **SilverScreen**. This involute profile has a base circle whose radius is equal to 20 and its center is at (0,0). The unwrapping process starts and ends at angles equal to 95 to 60°, respectively. Eleven points are generated using a constant angle decrement of 3.5°.

The program **Involute.PM** is listed below. A four-way, split-screen display (to be discussed in the next section) of the involute profile is shown in Figure 2.8.5.

---

[3]J. E. Shigley and L. D. Mitchell, *Mechanical Engineering Design*, 4th ed., McGraw-Hill, New York, 1983.

[4]Y, C. Pao, *Elements of Computer-Aided Design and Manufacturing, CAD/CAM*, John Wiley, New York, 1984.

Figure 2.8.4.

'F-Mesh Program Involute.PM — Draws Involute Gear Tooth
$C1 = 95*PI/180$
$C2 = 3.5*PI/180$
CN: 0 to $10*C2$ by $C2$
SN: -1 to 1 by 2
$X = (20*\cos(C1\text{-}CN)\text{-}20*CN*\cos(C1\text{-}CN\text{-}PI/2))*SN$
$Y = 20*\sin(C1\text{-}CN)\text{-}20*CN*\sin(C1\text{-}CN\text{-}PI/2)$

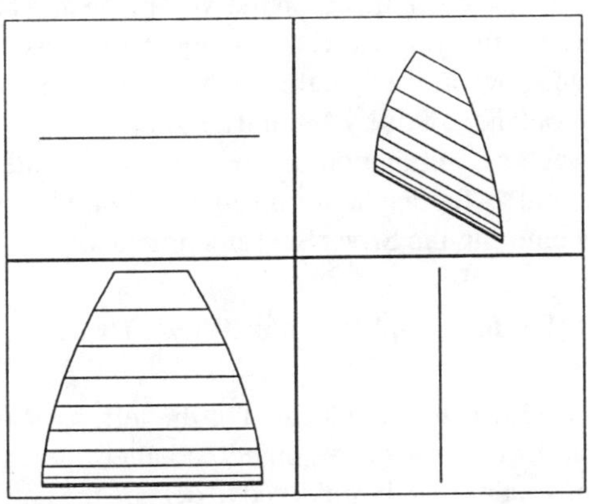

Figure 2.8.5.

2.9 SPLIT-SCREEN, MULTIPLE-WINDOW DISPLAYS

Figure 2.8.5 shows four different views of the involute profile generated with the **F-mesh** approach. A similar presentation is given in Figure 2.8.1 for the square mesh. To obtain such a printer hardcopy of what is being displayed on screen, we must first request that the screen be split into four different **windows** to show the top, front, isometric, and right-side views of the object at the upper left, lower left, upper right, and lower right regions of the screen, respectively. **SilverScreen** provides a **Window** menu shown in Figure 2.9.1 that enables users to select various types of screen splitting. In this text, we will introduce as many different types of multiple windows as necessary. Here, the *four-way, split-screen* arrangement is explained.

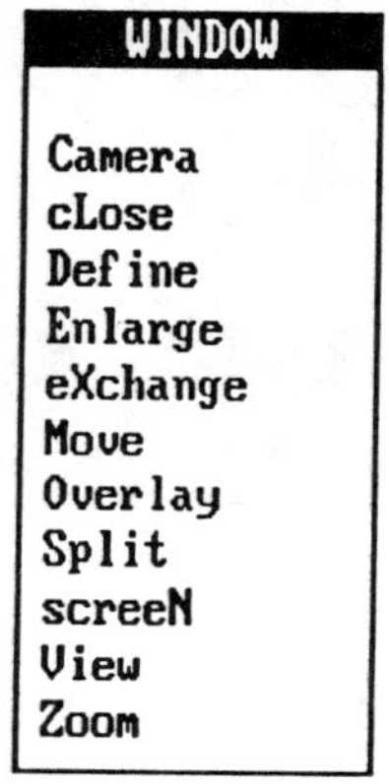

Figure 2.9.1.

So far, we have discussed how to make a two-dimensional drawing on the screen. If an object involves with vertices whose *Z*-coordinates are all equal to zero, this two-dimensional object then has been drawn on the *X–Y* plane and has no thickness. This *X–Y* plane is considered as the front view. If this object is viewed from the top and from the right side, it should show only a straight line because of zero thickness. The object may also be rotated about the *X*-axis by 45° and also about the *Y*-axis by 45° to produce an **isometric** view. Such a combination of front, top, right-side, and isometric view constitutes a fundamental set for displaying an object. Figure 2.8.5 is a typical example, which is obtained by entering the **SilverScreen** commands

Window Split Four-Way Drawing

Among the four windows, the one located at the left, lower region of the screen currently showing the front view of the object is called the **current window**. The drawing in the current window can be enlarged to occupy the entire **Graphics Area** of the screen by entering the commands **Window** and **Enlarge**. The drawings appearing in the other three windows can also be exchanged with the drawing in the current window by entering commands **Window** and **eXchange**. The window to be exchanged with the current window is to be specified using the pointing device or the arrow keys on the keyboard. Figure 2.9.2 illustrates the four-way, split-screen displays of the involute profile after the isometric view has been switched to the current window.

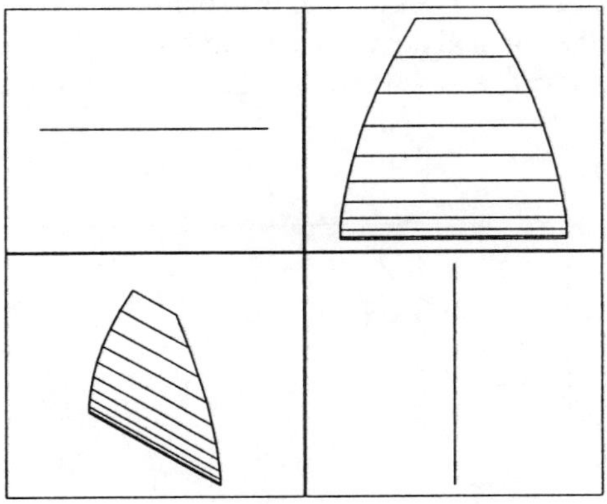

Figure 2.9.2.

The drawing in the current window can be enlarged by entering commands **Window** and **Enlarge**. The resulting display has a single window that shows the enlarged view of the object in the **Graphics Area** of the screen. Figure 2.9.3 is a typical example of the enlarged involute profile.

Another feature of **SilverScreen** in displaying a drawing is to zoom a selected portion for further enlargement. For example, if one wishes to enlarge the area defined

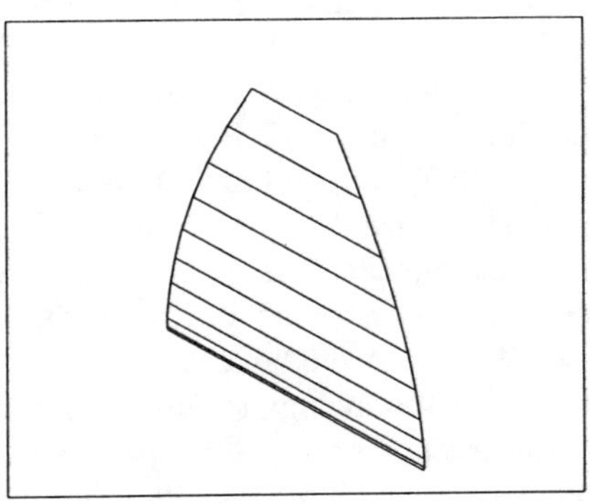

Figure 2.9.3.

by the broken lines (an add-on and not a part of the Enlarge operation) shown in Figure 2.9.3 the commands **Window**, **Zoom**, and **Area** can be entered to have the prompting messages

Mark first corner of area:
Mark opposite corner of area:

A pointing device such as a mouse is to be used to define the two required corners. The resulting drawing of the zoomed area is shown in Figure 2.9.4. Other features in the **Window** menu of **SilverScreen** will be introduced at appropriate time when some particular need arises.

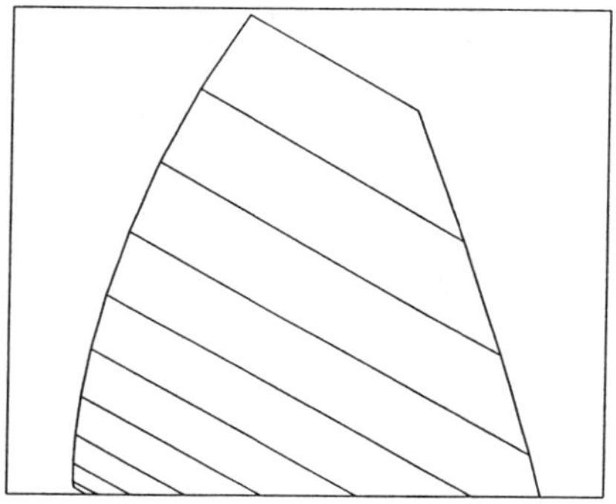

Figure 2.9.4.

2.10 CONCLUDING REMARKS

In this chapter, we have introduced some fundamental ways for drawing two-dimensional objects by application of **SilverScreen**. A drawing such as the thin plate with five holes shown in Figure 2.7.1, which has four sharp corners, may need to be modified by a process of using fillet or chamfer. Also, the dimensions of the plate and the five holes and the locations of the holes all must be clearly detailed. In the next chapter, we will discuss the methods of modifying a two-dimensional drawing. Figure 2.10.1 illustrates the detailing that needs to be completed; this is a topic covered in Chapter 4. Once a drawing is considered satisfactory and worthy of saving it as a drawing file on a disk, methods of storage and retrieval of the drawing and other types of files must be introduced. This is covered in Chapter 5.

Discussion of two-dimensional drawings is only the prelude of solid modeling. Starting from Chapter 6, construction of three-dimensional objects by application of **SilverScreen** will be addressed. What have been introduced in this chapter lays the foundation needed for the extended work of dealing with the real-world, three-dimensional objects. For instance, the **Window** menu of **SilverScreen** will be found

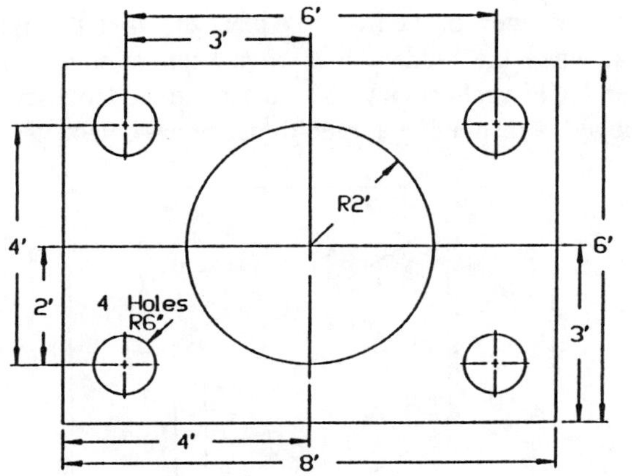

Figure 2.10.1.

most effective in providing the multiple-way, split-screen display of the object. In particular, the features of **isometric view** are important for bringing out the true characteristics of a three-dimensional object. This is demonstrated in Figure 2.10.2.

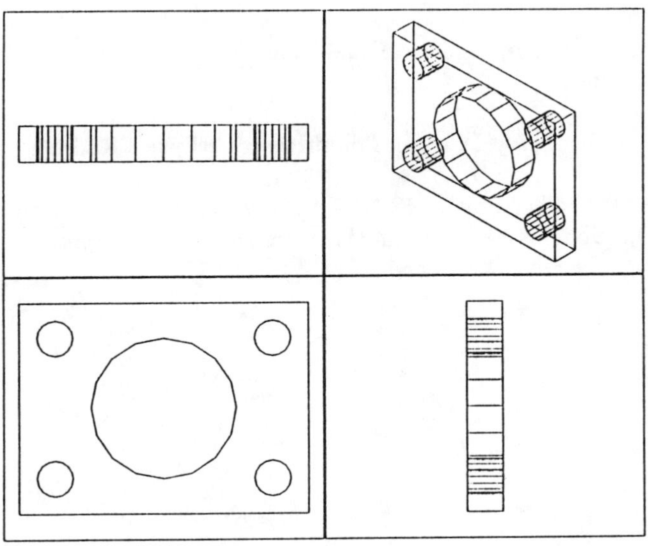

Figure 2.10.2.

The window exchange and enlarge methods introduced in this chapter will also be found important in application of the **hidden-line removal technique** for displaying the isometric view of a three-dimensional object. Figure 2.10.3 illustrates an enlarged version of the isometric view of the five-hole plate shown in the right, upper window of Figure 2.10.2, when the hidden lines have been removed. In Chapter 6, the **Presentation** menu of **SilverScreen** will be further utilized to discuss how the hidden lines can be removed in displaying a three-dimensional object.

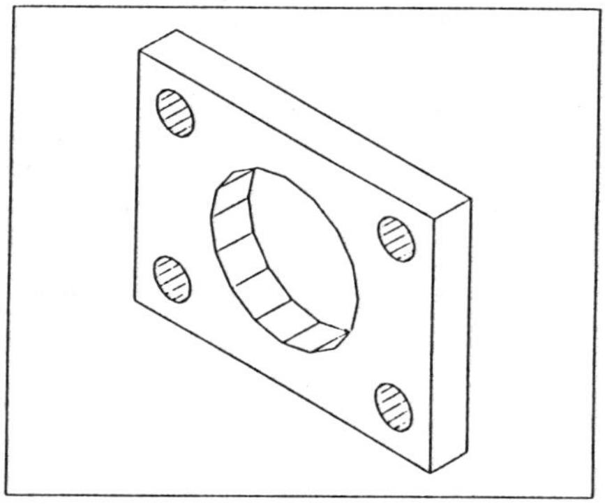

Figure 2.10.3.

PROBLEMS

2.1 Draw a circle of radius equal to 6″ and centered at $X = 4'$ and $Y = 8'$ and then apply the commands

Geometry Copy Primitives Cirlipse Radial

to generate a radial array of 12 circles of radius equal to 6″ as shown in Figure P2.1. These circles are 30° apart on the circumference of a circle of radius equal to 4′ and centered at $X = 8'$ and $Y = 8'$.

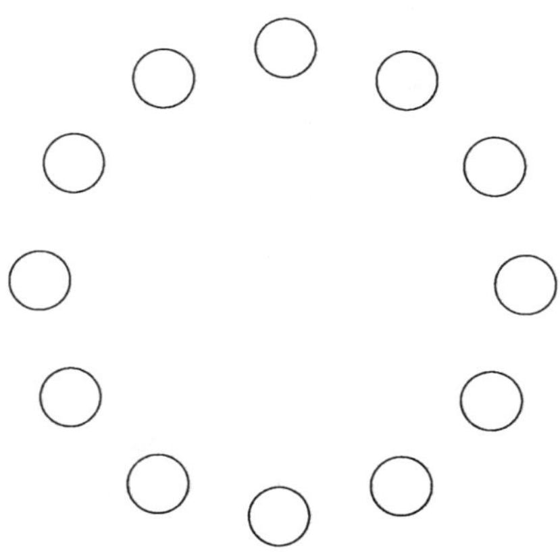

Figure P2.1.

2.2 Add center lines and apply **Copy** capability of **SilverScreen** to Figure P2.1 for all circles.
2.3 Draw one or more of the equilateral polygons shown in Figure 2.4.2.
2.4 Draw the framework for an arch window shown in Figure P2.4.

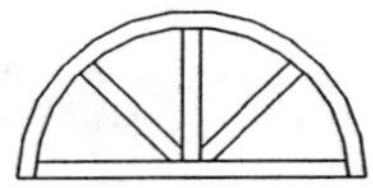

Figure P2.4.

2.5 Draw the extension and compression springs (Figure P2.5), with five and four active coils, respectively.
2.6 Draw a point at the left, lower corner of the rectangular border shown in Figure P2.6 and then apply the **Copy** and **Matrix** commands in the **Modify** menu of **SilverScreen** to generate a rectangular gridwork as illustrated.

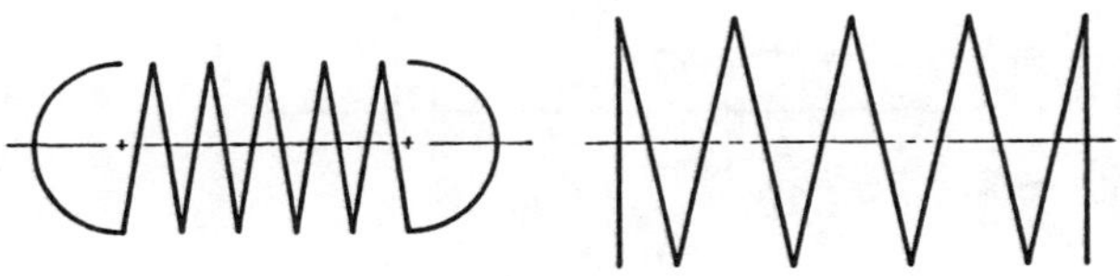

Figure P2.5.

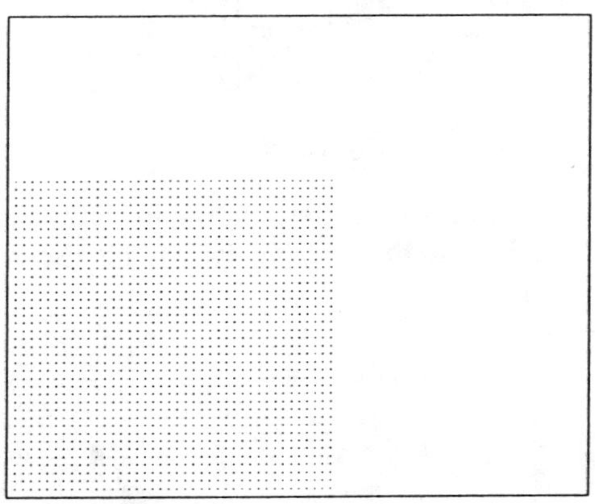

Figure P2.6.

2.7 Draw either one of the stove bolts shown in Figure P2.7.

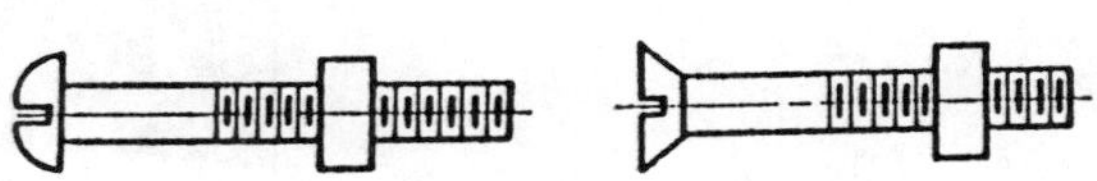

Figure P2.7.

2.8 Draw the button head machine bolt shown in Figure P2.8.

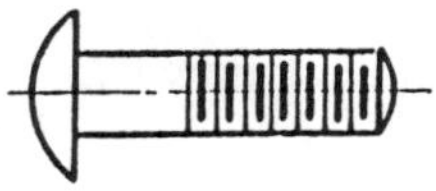

Figure P2.8.

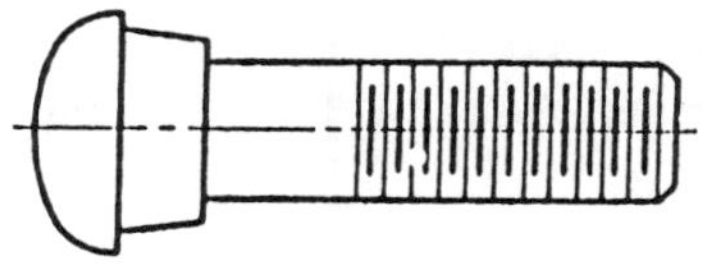

Figure P2.9.

2.9 Draw the oval neck track bolt shown in Figure P2.9.

2.10 Draw the two-dimensional schematic of a pipeline system shown in Figure P2.10.

2.11 Draw a two-dimensional outline of the electric bulb shown in Figure P2.11.

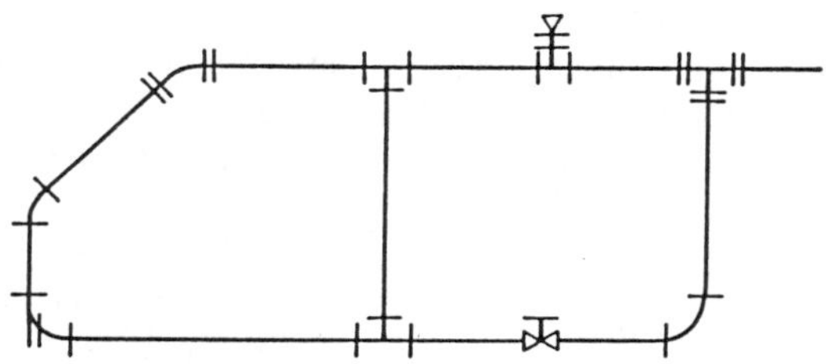

Figure P2.10.

Figure P2.11.

2.12 Draw the front views of the switch plates shown in Figure P2.2. Adding the labels "RESET" and "TEST" onto the switch plate will be discussed in Chapter 4.

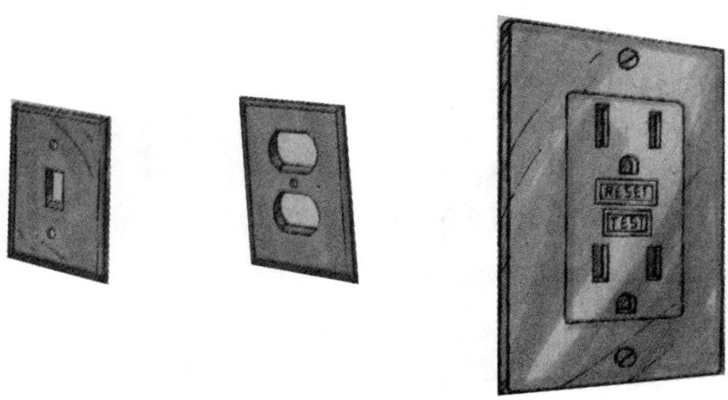

Figure P2.12.

2.13 Draw the two-dimensional model of a personal computer system shown in Figure P2.13.

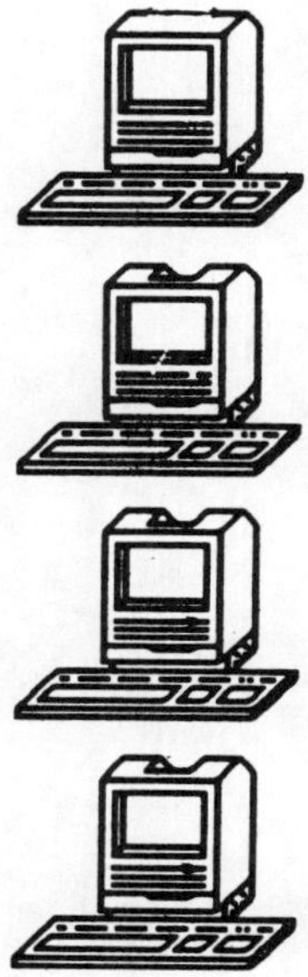

Figure P2.13.

2.14 Apply the procedure for obtaining four-way, split-screen displays of Figures 2.3.2 and 2.3.3. The results should be as shown in Figure P2.14.

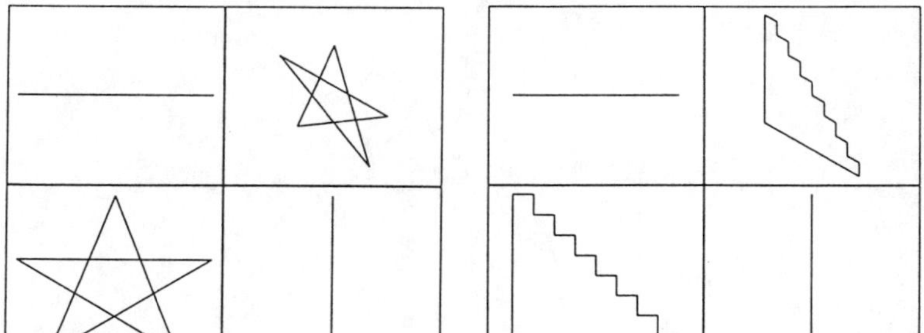

Figure P2.14.

3 Editing and Modifying Two-Dimensional Drawings

3.1 INTRODUCTION

In this chapter, we discuss how an existing drawing can be edited and modified by use of the **Edit** and **Modify** menus of **SilverScreen**, shown in Figure 3.1.1a and b, respectively.

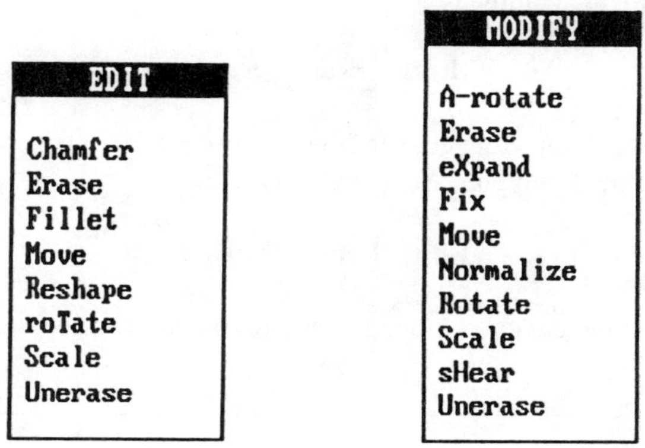

(a) (b)

Figure 3.1.1.

Since a two-dimensional drawing is constructed with points specified by the (X,Y,Z) coordinates, the drawing can be erased in part or entirely by deleting the lines that connect these specified points. **Erase** is an indispensible need in editing and modifying an existing drawing. **Unerase** is also needed in case an erase is incorrectly executed.

Also, a drawing can be relocated by changing the values of the coordinates of the points involved in that drawing. This is a simple **move** of the drawing by **translation**. **Rotation** about a desired axis should also be included in consideration of modifying an existing drawing.

The size of an object can be changed by using a uniform **scaling** factor for all three coordinate axes. Also, the shape of an object can be changed by using *different* scaling factors for the three coordinate axes, or by applying a **shear** operation.

In the following sections, we will introduce a number of **Edit** and **Modify** capabilities of **SilverScreen** and illustrate the operations with some practical examples.

3.2 ERASE AND UNERASE

To erase a drawing in part or in its entirety, the **Edit** menu of **SilverScreen** can be applied. For illustrating the procedures involved, let us load a drawing such as the F-mesh of an involute tooth-profile shown in Figure 2.8.4. This mesh consists a series of connected polygons (which can be examined by the **SilverScreen** commands **Structure**, **List**, and **Screen**). Suppose that we want to erase the polygon at the top. We enter the commands

Edit Erase Polygon

and then use a mouse to select the top polygon and press either the left button of the mouse or the **Enter** key. The resulting display is shown in Figure 3.2.1a. Same result can be achieved by the commands

Edit Erase Edge

if the cursor is moved to any four edges of the top polygon except the bottom edge since it is shared by the next polygon. Also, the commands

Edit Erase Vertex

will do the same if the cursor is moved to any one of the two top vertices.

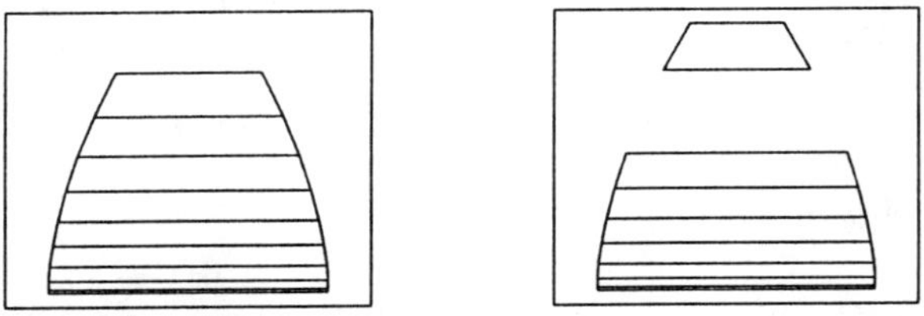

(a) (b)

Figure 3.2.1.

If the above Vertex method is used but the cursor happens to have been located at either one of the two end points of the second horizontal line on the top, both of the top two polygons will be erased as shown in Figure 3.2.1b. To obtain the desired outcome shown in Figure 3.2.1a from the reduced drawing Figure 3.2.1b, one may do the

unerase by using the command

Edit Unerase

and moving the mouse to select the second top polygon which needs to be recovered.

Fence command can be used to *fence in* a number of adjacent parts of a drawing for deletion or other modifications. For example, if Figure 3.2.1b is the desired, reduced drawing to be obtained from Figure 2.8.4, we enter the commands

Edit Erase Fence Inside

and use a mouse to define two corner points for defining a rectangular fence to enclose the two top polygons. Figure 3.2.2a shows the fenced area, indicated by the broken-line rectangle, and Figure 3.2.2b shows the identical result as in Figure 3.2.1b. All parts of the drawing outside of the defined fence may be erased by the commands

Edit Erase Fence oUtside-plus

and the resulting drawing is shown in Figure 3.2.2c. The command **oUtside-plus** erases *all* polygons outside of the defined fence while another command **Outside** requires erasing them by a repetitive Erase steps.

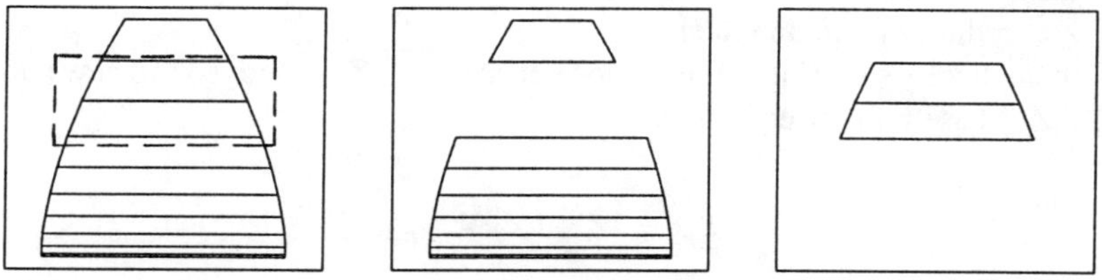

(a) (b) (c)

Figure 3.2.2.

If one wants to erase a number of polygons that are not connected, fence technique is then not workable. But we can put the polygons that are to be erased in a group by the commands

Edit Erase p-Group New

and use a mouse to continuously identify those polygons that ought to be included in the **new** group (or added into a **current** group). The **F10** key is to be pressed to end the selection. Figure 3.2.3 shows a typical result of the erase operation using the **p-Group** method.

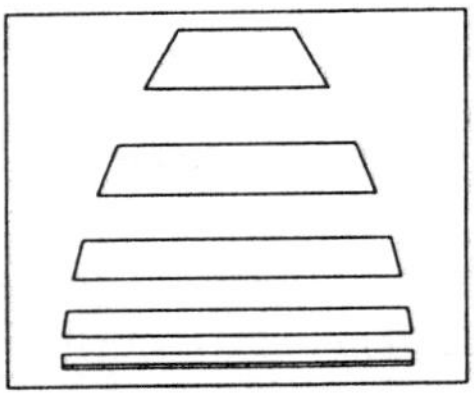

Figure 3.2.3.

3.3 TRANSLATION AND ROTATION

Translation of a drawing to a new location can be easily accomplished by the **Move** command of **SilverScreen**. For example, we may move the involute tooth-profile shown in Figure 2.8.4 to a new location by entering

Modify Move W-space
Mark or enter base point: 1,23
Mark or enter second point: 2,22

The whole drawing will be moved 1 unit to the right (X-axis) and 1 unit up (Y-axis). All points involved in constructing the drawing will have the values of the X coordinates increased by 1 and the values of the Y coordinates decreased by 1 unit, and the values of the Z coordinates remain intact.

Mathematically speaking, a point P located at (X_P,Y_P,Z_P) is moved to a new location (X_P',Y_P',Z_P') based on the equations

$$X_p' = X_p + (X_T - X_B) \quad (1)$$

$$Y_P' = Y_P + (Y_T - Y_B) \quad (2)$$

and

$$Z_P' = Z_P + (Z_T - Z_B) \quad (3)$$

where (X_T,Y_T,Z_T) and (X_B,Y_B,Z_B) are the coordinates of the *target* and *base* points, respectively, entered in the process of applying the **Move** command.

For rotating a drawing, the **A-rotate** and **Rotate commands** in the **Modify** menu of **SilverScreen** can be applied. A-rotate is to rotate a drawing about a defined axis. If no axis is specified, the rotation is about the Y-axis. For example, Figure 3.3.1a is copied from Figure 2.8.4 while Figure 3.3.1b is the result when the drawing is rotated about the Y-axis by 30° using the commands

Modify A-rotate
Enter angle of rotation: 30

The command **Rotate** differs from **A-rotate** in specifying the rotation with respect to a defined point. Figure 3.3.2 a and b illustrates the results of the following two respective commands:

Modify Rotate Drawing W-space
Enter angle(s) of rotation: x30

and

Modify Rotate Drawing W-space
Enter angle(s) of rotation: z30

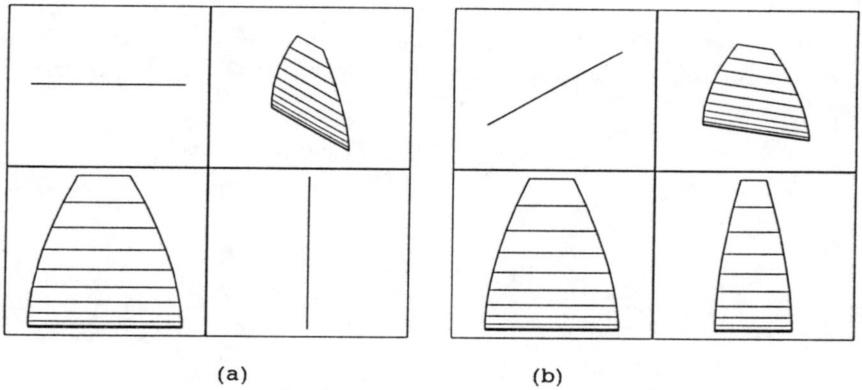

Figure 3.3.1.

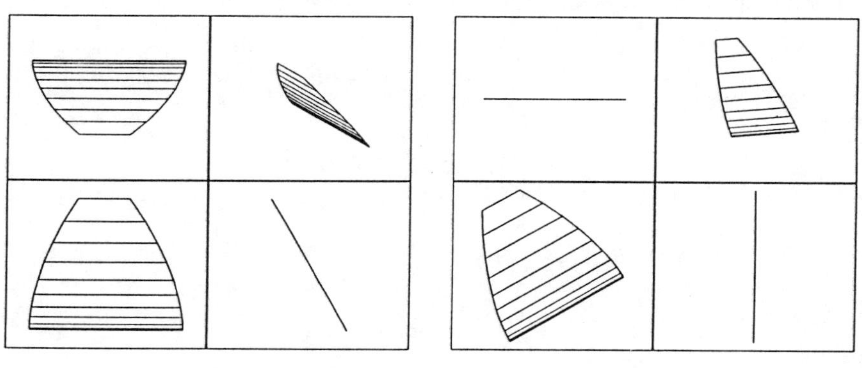

Figure 3.3.2.

Mathematically speaking, the rotation process transforms the coordinates of the vertices of the lines involved in the construction of an object. For a rotation about the *X*-axis by A_X radians, a point *P* located at (X_P, Y_P, Z_P) is moved to a new location with coordinates (X_P', Y_P', Z_P') based on the equations[1]

$$X_P' = Y_P \cos A_X + Z_P \sin A_X \tag{4}$$

$$Y_p' = -Y_p \sin A_X + Z_p \cos A_X \tag{5}$$

and

$$Z_p' = Z_p \tag{6}$$

Similarly, it can be shown that the equations for finding the coordinates of the new location (X_P', Y_P', Z_P') when the point P is rotated about the *Y*-axis by A_Y radians are

$$X_p' = X_p \cos A_Y - Z_p \sin A_Y \tag{7}$$

$$Y_p' = Y_p \tag{8}$$

and

$$Z_p' = X_p \sin A_Y + Z_p \cos A_Y \tag{9}$$

For a rotation about *Z*-axis by A_Z radians, the equations are

$$X_p' = X_P \cos A_Z + Y_P \sin A_Z \tag{10}$$

$$Y_p' = -X_P \cos A_Z + Y_P \sin A_Z \tag{11}$$

and

$$Z_p' = Z_P \tag{12}$$

---

[1]See, for example, Y. C. Pao, *Elements of Computer-Aided Design and Manufacturing*, CAD/CAM, John Wiley, New York, 1984.

As another example of applying the rotation and translation process, Figure 3.3.3 gives a demonstration of how some selected shapes and characters listed at the bottom of the screen can be copied, rotated if necessary, and then placed at a desired location in the construction of a chemical compound.[2] These selected shapes and characters are first created by use of the **Draw** and **Text** (to be discussed in the next chapter) commands of **SilverScreen**. A drawing file **ChemComp.DRW** (to be discussed in Chapter 5) is keeping all of the detailed steps, including the commands of **SilverScreen** that have been employed.

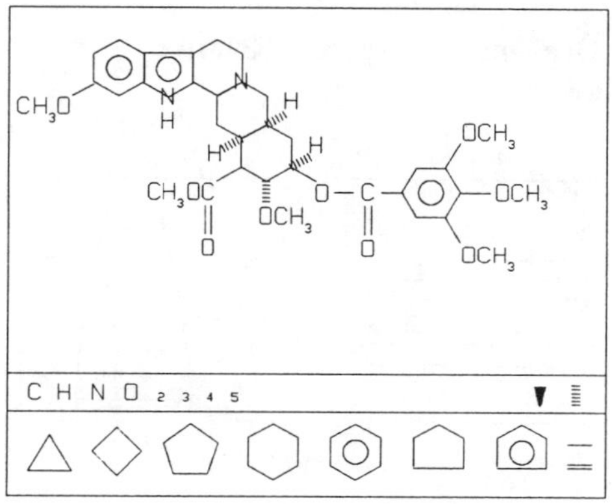

Figure 3.3.3.

3.4 SIZE AND SHAPE CHANGES

The size of an existing drawing can be scaled uniformly or nonuniformly in the *X*-, *Y*-, and *Z*-dimensions. For uniform scaling, the **shape** of the drawn entity will remain unchanged. For nonuniform scaling, the shape of the drawn entity will be distorted.

Figure 3.4.1a and b shows enlargement and reduction of the involute tooth-profile, respectively. They are the results of using the commands

Modify Scale Drawing W-space Absolute
Enter new dimension: 2,1,1

and

Modify Scale Drawing W-space Absolute
Enter new dimension: .5,1,1

[2]See, for example, J. L. Bentley, L. W. Jelinski, and B. W. Kernighan, "CHEM — A Program for Phototypesetting Chemical Structure Diagrams," *Comput. Chem.*, 11(4), 1987, 281–297.

The **Absolute** coordinates referred to the original data whereas the **Relative** coordinates refer to the currently displayed data. For example, if Figure 3.4.1b is currently shown on screen, we may obtain the original-size drawing as shown in the left, lower panel of Figure 3.3.1a by the commands

Modify Scale Drawing W-space Absolute
Enter new dimension: 1,1,1

or,

Modify Scale Drawing W-space Relative
Enter new dimension: 2,1,1

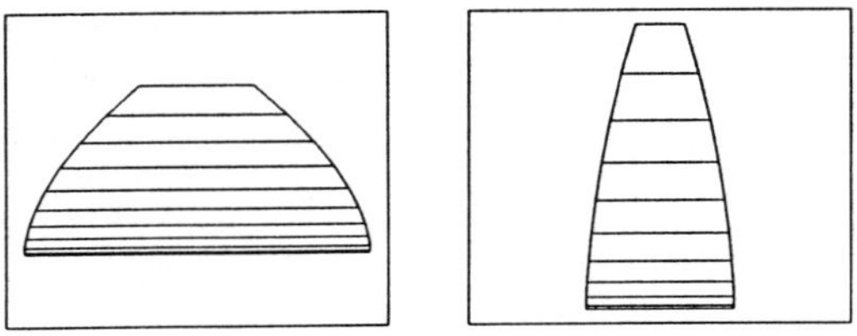

(a) (b)

Figure 3.4.1.

Shear command in the **Modify** menu of **SilverScreen** can be applied to obtain a *sheared* picture of an existing drawing. Let we use the involute tooth-profile drawing to demonstrate its usage. Figure 3.4.2a and b is the result of using the respective commands

Modify sHear Drawing W-space
Enter angle(s) of shear: 15

and

Modify sHear Drawing W-space
Enter angle(s) of shear: 45

Notice that both shears are referring to the z-axis and the counterclockwise direction is considered as positive. The limiting values for the shear angles are set in the **SilverScreen** software to the range of -85 to 85°.

As another example of applying the **sHare** command, we may draw a circle and shear it to generate an ellipse. Figure 3.4.3 shows that a circle is first drawn as an object

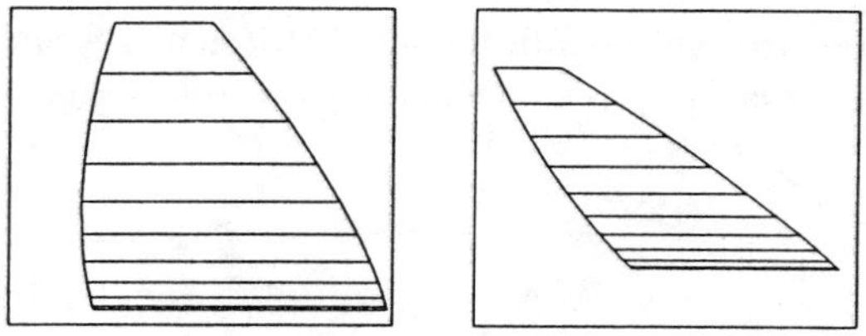

(a) (b)

Figure 3.4.2.

and placed on the left side of the screen and then copied once (selecting the **Single** option through the **Geometry** menu and the **Copy** command) and placed to the right side of the original circle. The ellipse shown on the right side of Figure 3.4.3 is generated by shearing the copied circle by an angle of 45°.

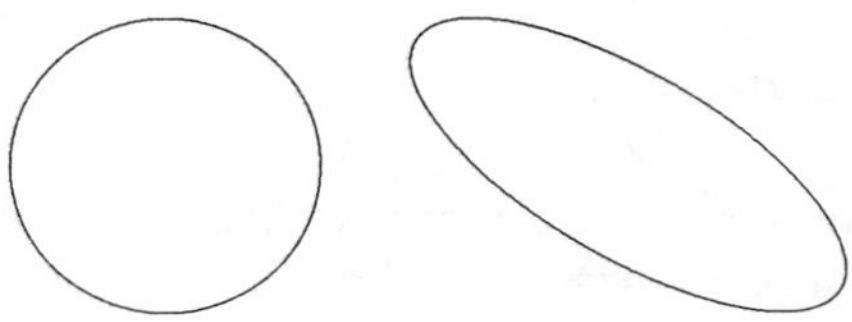

Figure 3.4.3.

3.5 MODIFYING CURVED LINES AND CONTOURS

In Section 2.7, we discussed how the **Bezier** command of **SilverScreen** can be effectively applied to draw curved lines and contours. Here, methods of altering such curved lines and contours are to be introduced for achieving any desired changes in their shapes and dimensions.

Consider the closed (top row) and open (bottom row) contours previously presented in Figure 2.7.4. In case some changes for these contours are to be made, we first retrieve the drawing by loading the drawing file (**Paof351.DRW**, referring to Chapter 5 for discussion of the drawing files). After we have seen the four contours on the screen, we enter the commands **Bezier**, **Edit**, and **reShape** to obtain a display as shown in Figure 3.5.1. It shows, in addition to the four contours, that there are circular and rectangular dots the points specified for drawing the contours, and *control* points, respectively. The left and right columns show *single* and *double* control points provided by the Bezier application and located between the specified points, respectively.

The coordinates of the specified and control points can be easily checked by moving the screen cursor to the point in question and reading the coordinates displayed in the

Status Area. These displayed coordinates also help identify where a newly specified point is during the reshaping process, which is to be discussed next.

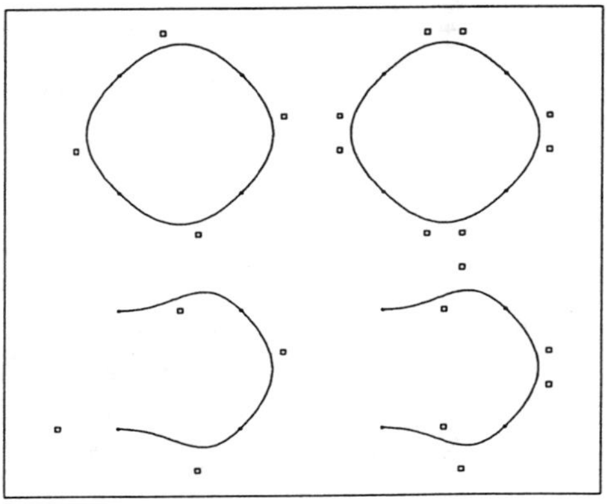

Figure 3.5.1.

To reshape the upper-left, closed contour shown in Figure 3.5.2a, we may select the point labeled 1 and change it to a new position. The outcome is exemplified in Figure 3.5.2b. Similarly, if the locations of the points labeled 2 and 3 are altered, the resulting, reshaped contours are shown in Figures 3.5.2c and d, respectively. For selection of a point for relocationing, **SilverScreen** provides the on-screen prompts: **F1 Next** point, **F2 Previous** point, and **F10 Finished**. When the **F10** key is pressed, a closed contour with a new shape is displayed.

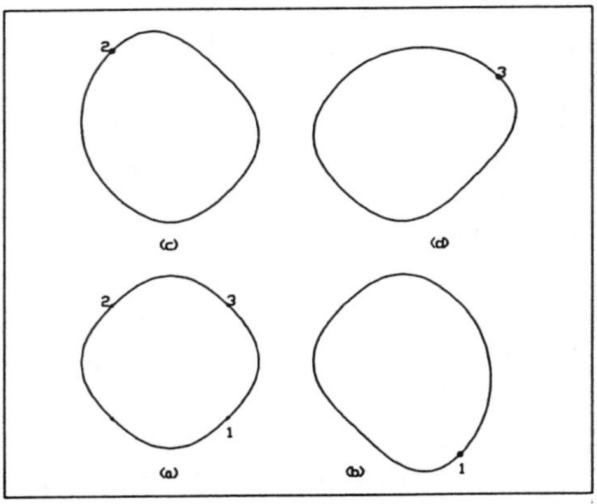

Figure 3.5.2.

Similar reshaping processes can be tried for the doubly controlled contour shown at the upper, right corner and also for both of the opened contours shown in the bottom row of Figure 3.5.1. This is left as exercises for the readers.

3.6 CONCLUDING REMARKS

There are many other useful commands in the **Edit** and **Modify** menus of **SilverScreen**. In this chapter, we have discussed only those commands that find most frequent uses. Users of **SilverScreen** are urged to work out the problems given at the end of this chapter to experiment the usage of **eXpand**, **Fix**, **Normalize**, and other commands.

Besides the **Edit** and **Modify** menus, other menus may provide helpful and should also be explored for editing and modifying any two-dimensional drawings and objects. For example, the **Structure menu** (Figure 3.6.1) has a **Join** command that can be applied to join two objects together. In the case of Figure 3.4.3, there are two objects, circle and ellipse. That is, each object contains only one primitive. We may prefer to combine them into a single object with two primitives.

```
STRUCTURE

Attribute
Create
Delete
Group
gOto
Join
List
Newname
Purge
Relocate
Tree
Undelete
```

Figure 3.6.1.

To know more about the geometric details of an object, the **Structure** menu and the commands **List**, **Object**, and **Screen** can be applied to have a list on the screen. Figure 3.6.2 is a printout of the list of a combined object consisting of two primitives, circle and ellipse, shown in Figure 3.4.3.

```
                              Silver List

OBJECT: obj1
  Base: 15',7'9",0'
  CIRCLE(1)  Center: 9',7'9",0'   Radius: 2  Surface Color: 15
  ELLIPSE(2)  Center: 15',7'9",0'  Surface Color: 15
```

Figure 3.6.1.

PROBLEMS

3.1 Draw Figure 3.4.3 by application of the **Circle** and **Ellipse** commands in the **Draw** menu of **SilverScreen**.

3.2 Draw Figure 3.4.3 by first drawing a circle and then copying it once and placing it on the right side of the original circle by use of the commands **Copy** and **Single** through the **Geometry** menu. The circle on the right is then sheared to become an ellipse.

3.3 Apply the **Polygon** command in the **Draw** menu to draw a six-sided hexagon and then apply the **Copy** command in the **Modify** menu to generate a two-dimensional "honeycomb" pattern.

3.4 Experiment reshaping with a doubly controlled, closed contour such as the one shown in the upper, right corner of Figure 3.5.1.

3.5 Experiment reshaping with a singly controlled, open contour such as the one shown in the lower, left corner of Figure 3.5.1.

3.6 Experiment reshaping with a doubly controlled, open contour such as the one shown in the lower, right corner of Figure 3.5.1.

3.7 Exercise the **Move** and **Scale** commands so that a large circle is reduced in size and can be placed touching an ellipse, which is generated by a shear operation of the circle, as is the case when Figure 3.4.3 is generated.

3.8 Generate the saw-tooth, cross-sectional view of the furnace filter shown in Figure P3.8 and then save the drawing in an object file. To learn how many primitives are involved, have a printout of them by application of the **List** command in the **Structure** menu.

3.9 Do the same as in Problem 3.8 but for a cylindrical, honeycomb air filter, of which a side view is shown in Figure P3.9.

Figure P3.8.

Figure P3.9.

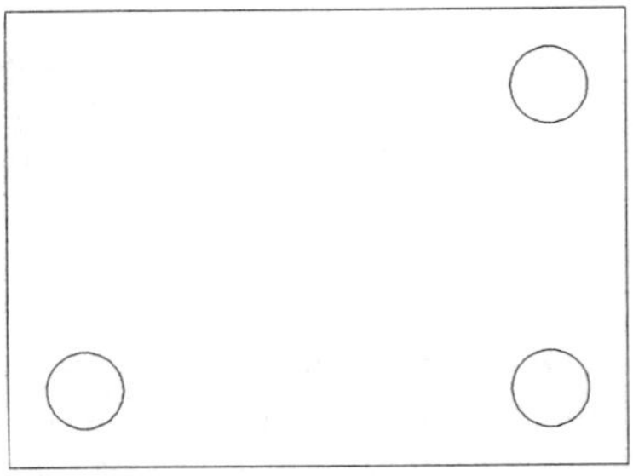

Figure P3.10.

3.10 Apply the **Erase** command in the **Edit** menu to erase the large circle at the center and also the small circle at the upper, left corner of the plate shown in Figure 2.4.1 so that the resulting display will be as shown in Figure P3.10.

4 Detailing Two–Dimensional Drawings

4.1 INTRODUCTION

In this chapter, we introduce the detailing capabilities of **SilverScreen**. Almost every two-dimensional drawing goes through a refining process of editing and reshaping of the lines and curves involved. After its shape and dimensions have been finalized, marking of its sizes with proper units and also addition of notes and remarks are then necessary to succinctly describe its geometrical characteristics. A circle may have been drawn using the commands in the **Draw** menu with the aid of a pointing device such as a mouse. But the location of its center needs to be indicated by addition of **center lines** and the size of the circle needs to be dimensioned by specifying its radius or diameter. If there are several circles involved in a two-dimensional drawing, the distances between any two circles will then have to be clearly marked.

Witness lines and **arrows** are necessary for pointing to which entity is being dimensioned. Also, a text line is necessary for giving a specific value or note. Version 1 of **SilverScreen** provides a **Text** menu for detailing of a drawing while Version 2 provides a new menu called **Annotation.** The commands available in these menus are shown in Figure 4.1.1a and b, respectively.

For adding fillets and chamfers, and for modifying already drawn dimensioning lines and text, the commands in the **Edit** menu of **SilverScreen** shown in Figure 4.1.2 can be applied. **SilverScreen** has a **Border** drawing that may be adopted to give a finishing touch to a completed drawing. As shown in Figure 4.1.3, the **Border** drawing arranges to have the lower, right corner for adding a title, marking the date when the drawing is made, and also the scale being used.

Other blocked areas could be added to the **Border** drawing for inclusion of items such as company name, draftsman's name, and material specifications. Other specially designed fonts such as company logo, and also notes could be added as shown in Figure 4.1.4. **SilverScreen** provides a number of standard font styles. Figure 4.1.4 shows two textstyles, **standard** and **italic**. In this chapter, we will discuss how to select different text styles and how to change the size of character and the spacing between characters, and other arrangements for adding a string of characters onto a drawing.

The user of **SilverScreen** may design a special font and include it in a textstyle library, such as in the case of a company's logo. This font can then be retrieved just like a standard character. In this chapter many examples will be given to delineate the procedures involved in various needs of detailing a drawing.

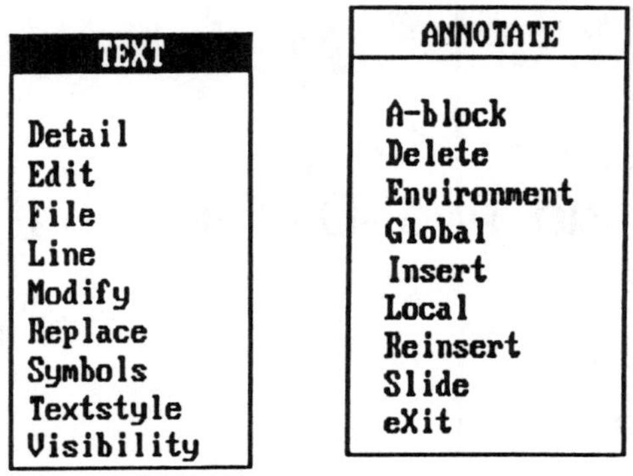

(a) (b)

Figure 4.1.1.

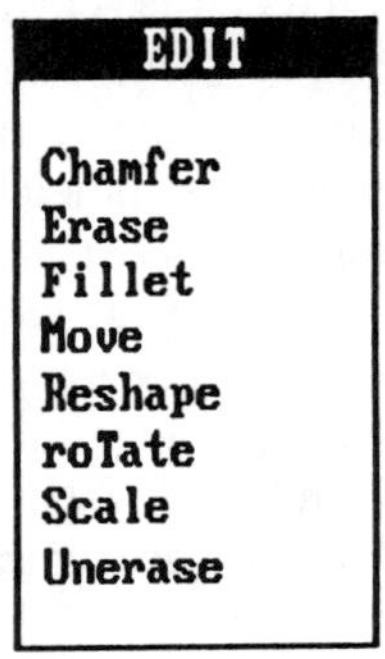

Figure 4.1.2.

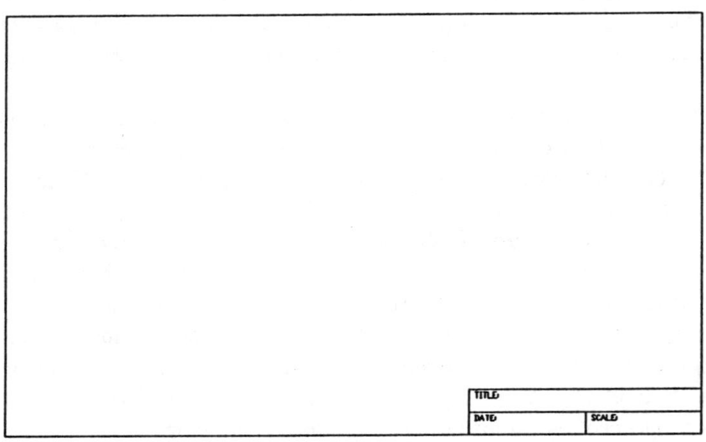

Figure 4.1.3.

This chapter addresses only the detailing of two-dimensional drawings. For dealing with three-dimensional objects, multiple two-dimensional detailings of the object's front, top, right side, and specially selected auxiliary views will then be necessary. Such needs will be discussed after we have introduced the methods for the construction of three-dimensional objects by application of **SilverScreen**.

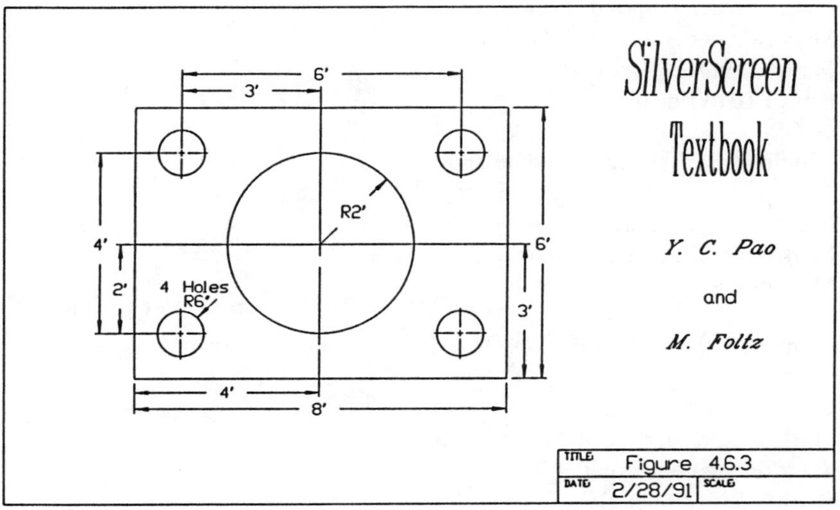

Figure 4.1.4.

4.2 CENTER LINES

To draw the center lines of a circle, we may first draw a + at the center and then one radial line such as the one shown in Figure 4.2.1. The **Copy** command in the **Geometry** menu can next be applied to radially copy this radial line to complete the center lines as shown in Figure 4.2.2. To discuss the dimensioning of a two-dimensional drawing in the ensuing sections and also the drawing of the center lines for the circle here, let us generate Figure 4.2.2 from the very beginning.

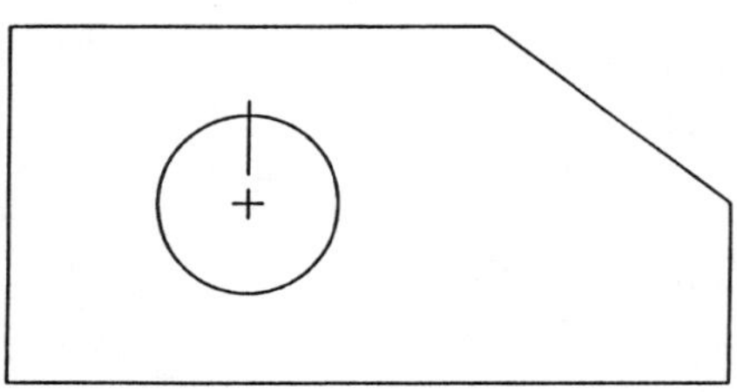

Figure 4.2.1.

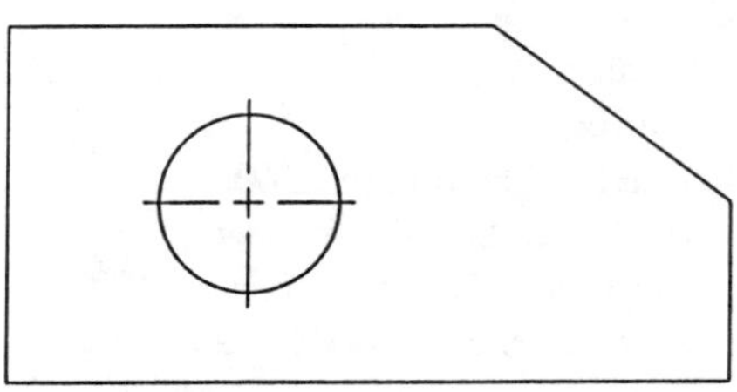

Figure 4.2.2.

The outside, closed contour is drawn with the commands

Draw Lines Multiple
Mark or enter start of first line: 0′,0′
Mark or enter endpoint: 0′,6′
Mark or enter endpoint: 8′,6′ (all Z = 0)
Mark or enter endpoint: 12′,3′
Mark or enter endpoint: 12′,0′
Mark or enter endpoint: 0′,0′
Press **F10** to finish

The circle is drawn using the commands

Draw Circle Radius
Mark or enter center of circle: 4′,3′
Mark or enter radius point: 5′6″,3′

The + mark at the center of the circle is drawn with the commands

Draw Lines Single
Mark or enter start of first line: 3′9″,3′
Mark or enter end point: 4′3″,3′
Draw Lines Single
Mark or enter start of first line: 4′,2′9″
Mark or enter end point: 4′,3′3″

The vertical center line is drawn with the commands

Draw Lines Single
Mark or enter start of first line: 4′,3′6″
Mark or enter end point: 4′,4′9″

The other three center lines are generated using the commands

Geometry Copy Primitives
(Use the pointing device to select the vertical center line)
Radial
Enter angle of copy: 360
Enter number of copies: 4
Mark or enter top of axis: 4′,3′,0′
Mark or enter bottom of axis: 4′,3′,1′

As another example, there are one large circle and four small circles for the plate previously presented in Figure 2.4.1. The above-described procedure for drawing the center lines can be followed for the large circle and one of the small circles to obtain the resulting drawing as shown in Figure 4.2.3. To add the center lines for the other three small circles, we can apply the **feNce** command to define a group of primitives

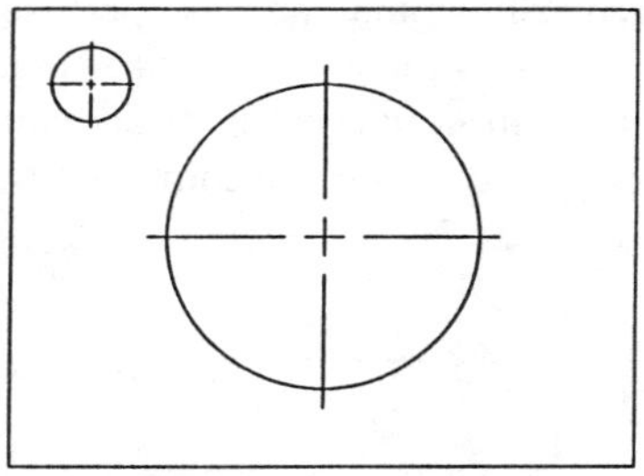

Figure 4.2.3.

fenced inside of a rectangular region for the purpose of copying it to the other specified locations. The commands involved in the fence and copy operations are

Geometry Copy Primitives feNce Inside
Mark first corner of area
Mark opposite corner of area

The pointing device or arrow keys on the keyboard are to be used to define the two opposite corners of the fence and the **Enter** key is to be pressed to confirm the decisions. The rectangular, fenced area for the operation of copying the small circle is illustrated in Figure 4.2.4a. The actual copying process follows the procedure previously discussed in Section 2.4. The matrix option should be selected in the copying process, and the fenced group of primitives including the small circle and its center lines should be copied to the other three specified locations. Two rows and two columns are to be specified and it is advisible to select the center of the small circle as the base point and the desired locations for the center of the small circles at the lower, left corner and upper, right corner as the target points. The resulting drawing after copying the fenced group is shown in Figure 4.2.4b.

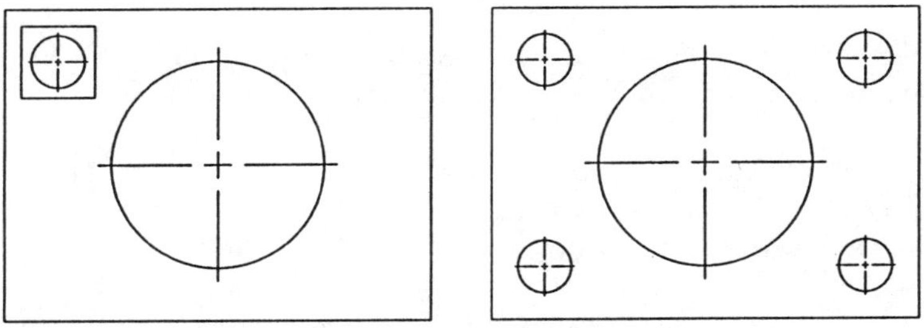

(a) (b)

Figure 4.2.4.

4.3 DIMENSIONING

In this section, we will discuss various capabilities of **SilverScreen** that can be applied for dimensioning. Consider the drawing shown in Figure 4.2.2. It indicates that a rectangular plate has a circular hole and has its upper, right corner chamferred. Its boundaries consist of horizontal, vertical, and inclined lines. For dimensioning, the lengths of these lines and the location and size of the circular hole need to be specified.

To detail this two-dimensional drawing, the command **Detail** in the **Text** menu can be applied. Figure 4.3.1 illustrates the dimensioning of a horizontal line, an inclined

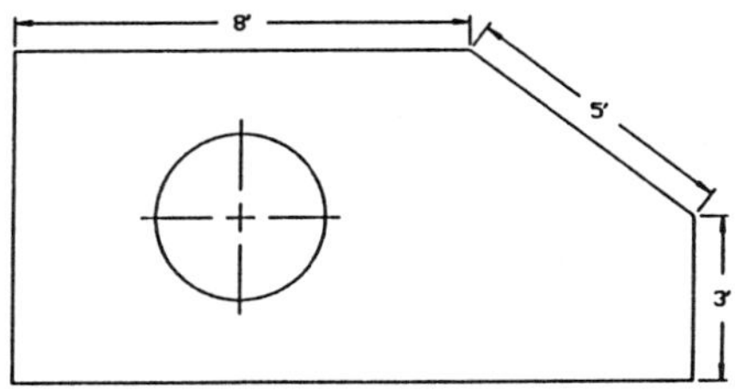

Figure 4.3.1.

line, and a vertical line. These three types of dimensioning are accomplished by applying the commands

Text Detail Insert Linear

and then selecting the **Horizontal**, **Parallel**, and **Vertical** options, respectively. As for which line the dimensioning should be made, the pointing device is to be used to move the screen cursor to the desired primitive.

For dimensioning the circle shown in Figure 4.3.2, we use the commands

Text Detail Insert Radius Rotated
Enter angle of rotation: 45
Select arc or cirlipse: (select the only circle)

In case we want the text to be horizontally oriented as shown in Figure 4.3.3, we may enter the commands for modification as follows:

Text Detail Modify Detail
Select detail: (select the radius annotation)
Text Orientation Horizontal

By adding the lengths for the left and bottom boundaries and also for the location of the circular hole, the dimensioning task is completed as shown in Figure 4.3.4. However, it should be mentioned that during the dimensioning process for the left

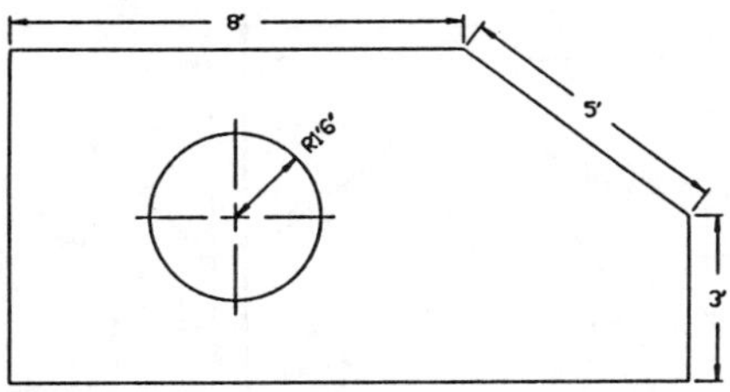

Figure 4.3.2.

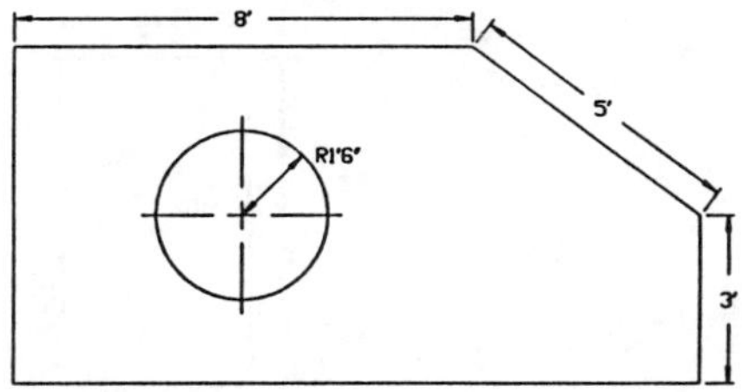

Figure 4.3.3.

boundary using the **Insert**, **Linear**, and **Vertical** commands, the annotation will appear on the right side of the left boundary line. The pointing device has to be used to relocate the annotation to a desired position and the **Enter** key be pressed to finalize the decision. This maneuver should also be applied for labeling the bottom edge and the location of the circular hole.

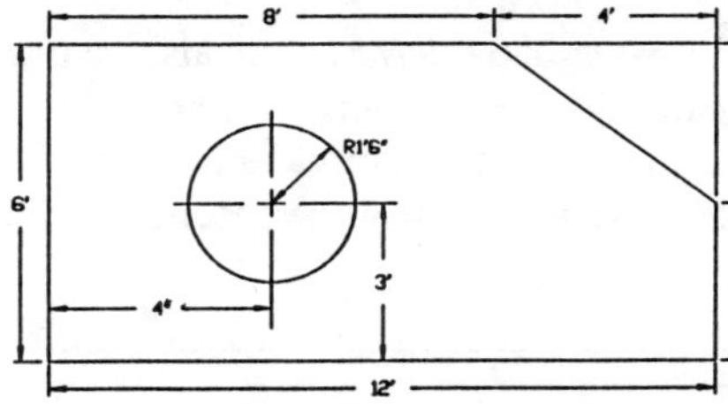

Figure 4.3.4.

SilverScreen has ways of modifying the drawing of the **witness** lines and **arrows** used in the annotation. As shown in Figure 4.3.5, the detailing of the left and bottom boundaries is not properly done. The **extension** and **offset** of the witness lines can be adjusted by the commands

Text Detail Modify Detail Witness

and then by selecting the options of **Extension** and **Offset**, respectively. The respective, follow-up questions that need to be answered are

Enter extension length:

and

Enter offset length:

By experimenting with different lengths, the desired witness lines can be achieved.

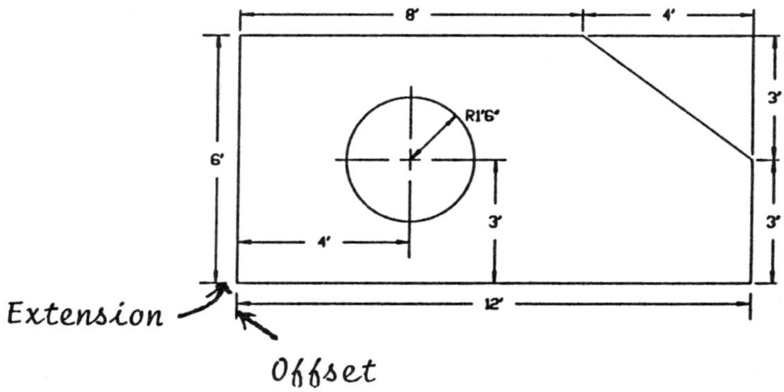

Figure 4.3.5.

TOLERANCE

In dimensioning the size of an object, some dimensions are permitted to vary within certain specified limits. This so-called *tolerance* can also be marked on a drawing by applying the detailing capability of **SilverScreen**. For example, Figure 4.3.6 is the result of modifying the dimensioning of the circular hole shown in Figure 4.3.5 if the radius is permitted to be 1″ more or less than the specified value.

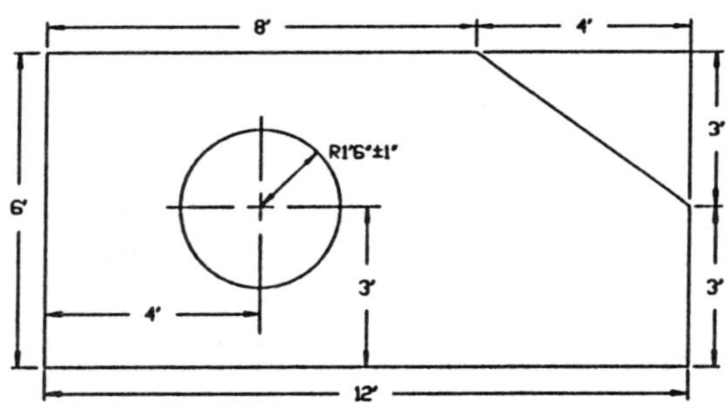

Figure 4.3.6.

The commands required to acheive this are

Text Detail Modify Detail Tolerance
Enter tolerance values (positive, negative): +1″,–1″

There are other options for dimensioning such as using only one arrowed line, changing a text to a different size, or using a leader arrow. We will introduce them and demonstrate with many examples in the ensuing sections.

4.4 FILLET AND CHAMFER

Fillet is a common practice for rounding the sharp corners of an object to prevent fracture. The thinner the object the greater the radius of the fillet should be for the sake of appearance and liability. Alternatively, chamferring is another way of cutting off a sharp corner or edge of an object without rounding it. **SilverScreen** has commands in the **Edit** menu for modifying a sharp corner of an object by the fillet or chamfer method.

Figure 4.4.1 shows that Figure 4.3.6 has been further modified by having a fillet of radius equal to 4″ made at the upper, left corner. This is accomplished by entering the commands

Edit Fillet Insert Lines
Select the first line
(Use the arrow keys or the pointing device to select the top, horizontal line and then press **Enter** key)
Select the second line
(Use the arrow keys or the pointing device to select the leftmost, vertical line and then press **Enter** key)
Enter radius of fillet: 4″

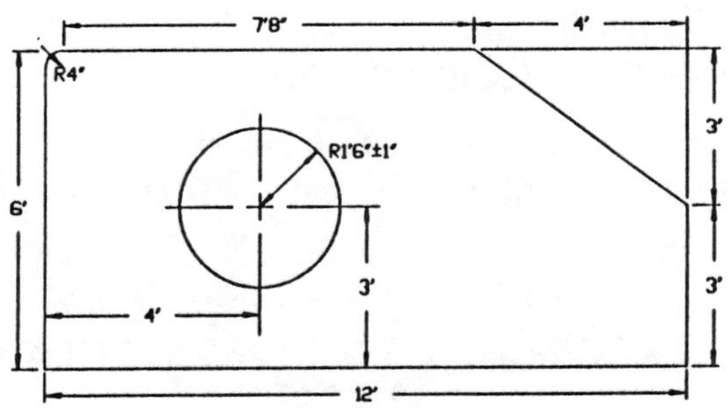

Figure 4.4.1.

The fillet can next be dimensioned using commands as have been explained in the preceding section.

Figure 4.4.2 is an example of chamferring the right, lower corner of the object shown in Figure 4.4.1. The commands required are

Edit Chamfer Insert Lines
Select first line
(Use the arrow keys or the pointing device to select the bottom, horizontal line and then press **Enter** key)

Enter first length: 2′
Select second line
(Use the arrow keys or the pointing device to select the top, horizontal line and then press **Enter** key)
Enter second length: 2′

The chamfer will be drawn and automatically dimensioned by **SilverScreen**. However, modification may sometimes be necessary for making them compatible with the others dimensionings already on the drawing.

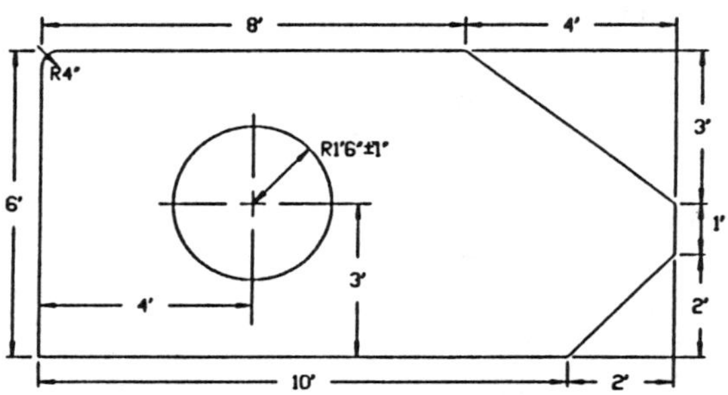

Figure 4.4.2.

4.5 SELECTING FONT AND ADDING TEXT

For adding text onto a drawing, **SilverScreen** provides four standard fonts: **Italic**, **Modern1**, **Roman1**, and **Txt**. Figure 4.5.1 lists samples of these four different font styles.

Italic	*abcdefghijklmnopqrstuvwxyz* *ABCDFGHIJKLMNOPQRSTUVWXYZ* *0123456789*
Modern1	abcdefghiJklmnopqrstuvwxyz ABCDFGHIJKLMNOPQRSTUVWXYZ 0123456789
Roman1	abcdefghijklmnopqrstuvwxyz ABCDFGHIJKLMNOPQRSTUVWXYZ 0123456789
Txt	abcdefghijklmnopqrstuvwxyz ABCDFGHIJKLMNOPQRSTUVWXYZ 0123456789

Figure 4.5.1.

To request a particular font selection, the commands to be entered are **I/O**, **Load**, and **Font**. The prompted message is

Enter font name:

The pointing device can be used to make the selection of the preferred one among the available four font styles. Alternatively, the name of the selected style can be entered via the keyboard.

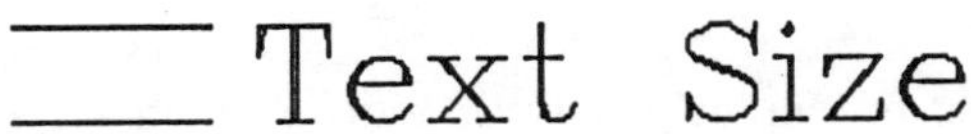

Figure 4.5.2.

A selected font style can be edited and new font styles can be created. In both cases, we need to define the **height** (Figure 4.5.2) and the **aspect ratio,** which is the ratio of width to height of the character. Figure 4.5.3 illustrates the appearance of a string of characters when the aspect ratios are specified with the values equal to 0.5, 1, and 2 when the height remains the same. Also, a **slanting angle** in the range from –85 to 85° can be specified. Figure 4.5.4 shows the cases of the string of characters "Slant" slanted by the angles of –20, 0, and 20°.

Figure 4.5.3.

Figure 4.5.4.

For handling a string of characters, the distance between adjacent characters may also be changed. Figure 4.5.5 shows cases where the spacing of the characters in the string "Spacing" has been specified with the values of 0.5, 1.0, and 2.0.

Figure 4.5.5.

The specifications of font, height, aspect ratio, slanting angle, and spacing of the characters in a string are to be entered by use of the commands **Textstyle** and **Define** in the **Text** menu of **SilverScreen**. Once a textstyle is defined, it will be given a name and can be exhibited by entering the commands **Text**, **Textstyle**, and **Exhibit**.

ADD TEXT ONTO DRAWINGS

The command **Line** in the **Text** menu of **SilverScreen** can be applied for placing a character or a string of characters at a selected location on a completed drawing. A series of questions are to be answered for specifying the content of the string, the character size, the spacing between adjacent characters in the string, the displaying angle of the string, and so on. Let us illustrate the procedure involved in producing the various text lines appearing in Figure 4.5.6.

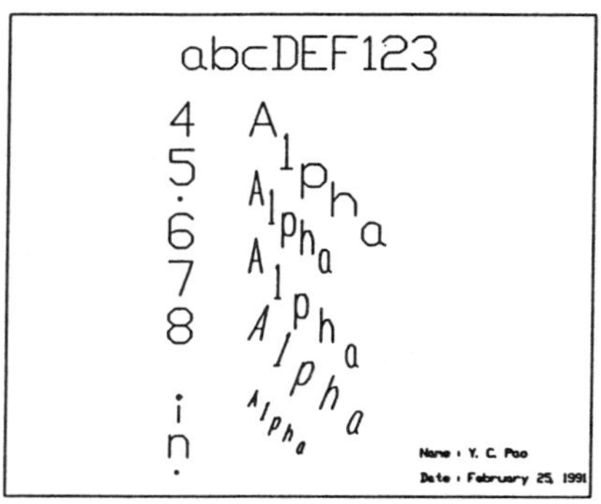

Figure 4.5.6.

TEXT LINE “abcDEF123”

In order to place a string of characters, abcDEF123, horizontally and having the lower, left corner of the leading character “a” at the location (3′,12′,0′), we enter the commands

Text Line
Enter text: abcDEF123
Standard
Enter insertion point: 3′,12′,0′

By requesting the standard font style, the character fonts of the **TXT** style listed in Figure 4.5.1 will automatically be provided for drawing text line.

TEXT LINE "45.678 in."

Text line may also be placed vertically. The vertical text line of 45.678 in. appearing in Figure 4.5.6 is the result of entering the following commands:

Text Line
Enter text: 45.678 in.
Textstyle
Enter name of textstyle: standard
Vertical Below
Mark or enter insertion point: 3′,11′,0′

The text line is placed **below** the point. There are also the **above** and **center** options for placing the text line above and centered equally on each side of the insertion point, respectively.

TEXT LINES "ALPHA"

Text line may also be drawn diagonally at a specified angle, which is considered as positive if measured **counterclockwise**. We shall refer the five, differently drawn "Alpha" in Figure 4.5.6 as #1 through #5 counting from the top and explain the procedure required for drawing them in detail below.

The uppermost, #1, text line "Alpha" in Figure 4.5.6 is drawn using the commands

Text Line
Enter Line: Alpha
Text Textstyle
Enter name of textstyle: standard
Diagonal
Enter angle: –45
LEFT Left-justified text
Mark or enter insertion point: 5′,10′,0′

The **LEFT-justified** request places the lower, left corner of the first character in the diagonally oriented text line at the insertion point. There are also the **RIGHT** and **CENTER** options, which place the lower, right corner of the last character in the text line at the insertion point and center the text line equally at each side of the insertion point, respectively.

The #2 text line of "Alpha" in Figure 4.5.6 is the result of adjusting the **aspect ratio** from 1 to 0.5. The commands involved are

Text Textstyle Standard Aspect
Enter new aspect ratio: 0.5
Enter Line: Alpha
Text Textstyle
Enter name of textstyle: standard
Diagonal
Enter angle: –45
LEFT Left-justified text
Mark or enter insertion point: 5′,8′,0′

The #3 text line of "Alpha" in Figure 4.5.6 is the result of adjusting the **spacing** from 1 to 2. The commands involved are

Text Textstyle Standard Spacing
Enter new spacing: 2
Text Line
Enter Line: Alpha
Text Textstyle
Enter name of textstyle: standard
Diagonal
Enter angle: –45
LEFT Left-justified text
Mark or enter insertion point: 5′,6′,0′

Notice that the aspect ratio remains equal to 0.5. Also, spacing will remain equal to 2 hereon until a new value is defined.

The #4 text line of "Alpha" in Figure 4.5.6 is the result of **slanting** the text line by an angle of 20°. The commands involved are

Text Textstyle Standard sLant
Enter new angle of slant: 20
Text Line
Enter Line: Alpha
Text Textstyle
Enter name of textstyle: standard
Diagonal
Enter angle: –45
LEFT Left-justified text
Mark or enter insertion point: 5′,4′,0′

The #5 text line of "Alpha" in Figure 4.5.6 is obtained when the **height** of the text line is changed from 1 to 0.5. The commands involved are

Text Textstyle Standard Height
Enter new height for character: 0.5
Text Line
Enter Line: Alpha
Text Textstyle
Enter name of textstyle: standard
Diagonal
Enter angle: –45
LEFT Left-justified text
Mark or enter insertion point: 5′,2′,0′

The current definition of the textstyle can be viewed by the commands

Text Textstyle Define

Figure 4.5.7 shows a typical status-display of the current setup of the text style. ⇒ is pointing to the Font: **txt**, which indicates that the standard textstyle is selected. A new style can be entered through keyboard input and the **Enter** key is pressed to confirm the change. Arrow keys on the keyboard or a pointing device such as a mouse can be used to move the ⇒ to the immediate left of the other specifications (Height, Spacing, Slant, and Aspect) and to enter new values. The **F10** key is pressed to finish the new definition of textstyle. Figure 4.5.8 is another example of defining the **italic** font with a normal character height and aspect ratio and a normal spacing between adjacent characters but with a slanting angle of 20°.

Figure 4.5.7.

Using the textstyle definition of Figure 4.5.7, additional text lines of “Name: Y. C. Pao” at an insertion point (11.75″,0.5″,0″) and “Date: February 25, 1991” at an insertion point (11.75″,0″,0″) are added to Figure 4.5.6 to obtain a finished text drawing. Notice that the height is indeed reduced to 0.25.

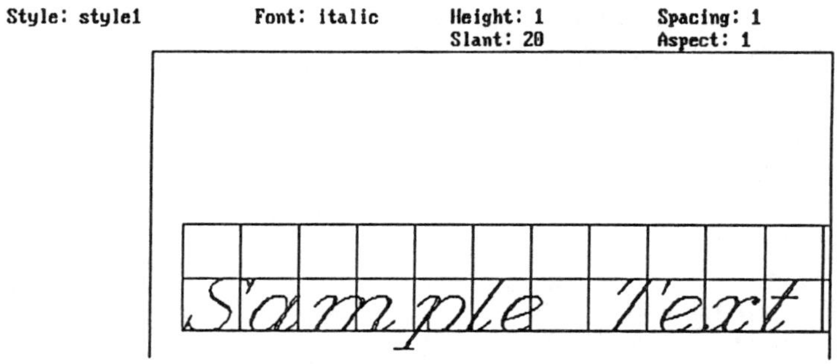

Figure 4.5.8.

In the next section, we discuss a formal way of completing a drafting task.

4.6 FINISHING TOUCH

After a two-dimensional drawing has been completely dimensioned such as the plate with one large hole and four small holes shown in Figure 4.6.1, one may wish to add some finishing touches to the drawing by adding name, date, title of the drawing, and other explanatory remarks. **SilverScreen** has a **Border** drawing (Figure 4.6.2) that can be loaded and then combined with the finished drawing. Additional decorative line and curves such as a company logo and text lines such as the institution or company name can certainly be detailed onto the drawing before a hardcopy is made.

Figure 4.6.3 illustrates a finished two-dimensional drawing after more text lines, title, and date have been added. It demonstrates the use of italic and standard text styles and additional adjustment in the slanting angle for the text line "SilverScreen." Size adjustments and relocation of text lines have also been exercised in producing this drawing.

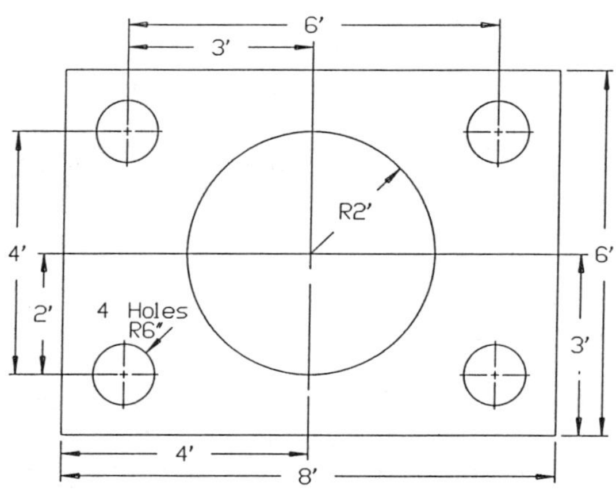

Figure 4.6.1.

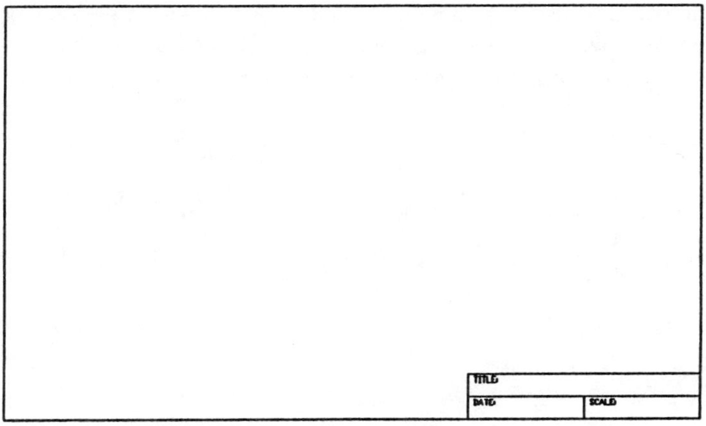

Figure 4.6.2.

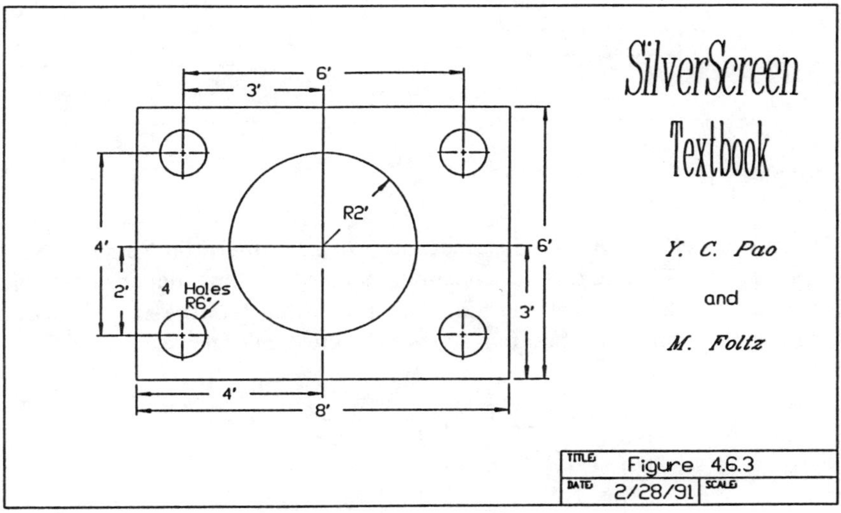

Figure 4.6.3.

4.7 CONCLUDING REMARKS

After having explained in this chapter how a two-dimensional drawing is to be detailed, we can extend the methodology to any three-dimensional drawings and objects that will appear in the ensuing chapters. The window-exchange capability of **SilverScreen** enables any two-dimensional drawing in the current window to be detailed. To detail a three-dimensional object, we may handle one-by-one its front, top, right-side, and also auxiliary views (to be discussed in Chapter 8) if there are any constructed. These detailed views can be selected, combined, and then put onto a multiple-window screen for eventual hardcopying either on a printer, or on a plotter. Figure 4.7.1 illustrates a typical example of such a three-dimensional detailing.

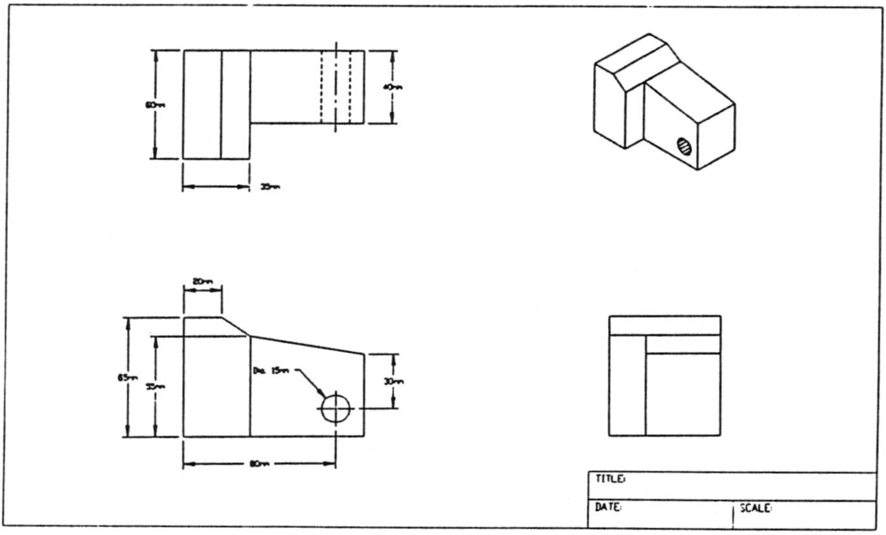

Figure 4.7.1.

Also, in Chapter 8 we will introduce the concept of the **Constrution Space (C-Space)** to facilitate the addition of text lines on different faces of a three-dimensional object. Figure 4.7.2 is an illustration of the addition of different labels on the six faces of a cube either by application of standard font styles or by creation of new styles.

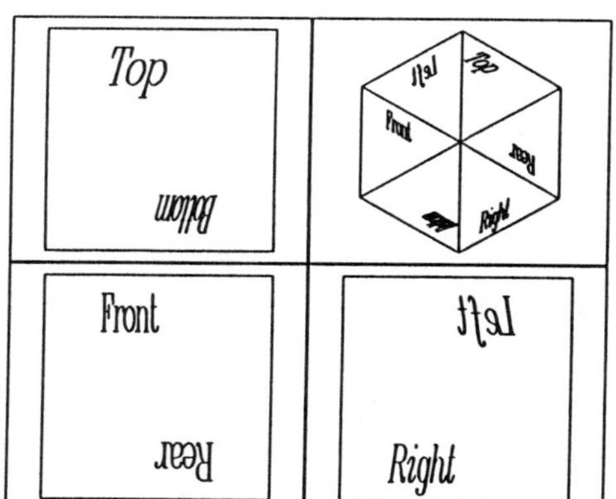

Figure 4.7.2.

Many other interesting topics can be introduced in application of **SilverScreen** for detailing of two-dimensional and three-dimensional drawings, such as the topic of how to save a newly created font into a font library by applying the **Library command** of **SilverScreen**. But to keep this text at an introductory level, we will leave it to the readers to explore the advanced topics and application by themselves through careful study of the **SilverScreen** users' menu.

PROBLEMS

4.1 After the fillet has been added into Figure 4.4.1, which commands should be used to add the radial dimensioning of R4″?

4.2 Implement the dimensioning of the chamfer shown at the lower, right corner of the drawing in Figure 4.4.2.

4.3 Use the proper commands of **SilverScreen** to reduce the size of the single arrow for dimensioning of the fillet shown in Figure 4.4.2 so that the resulting fillet dimensioning will appear as in Figure P4.3.

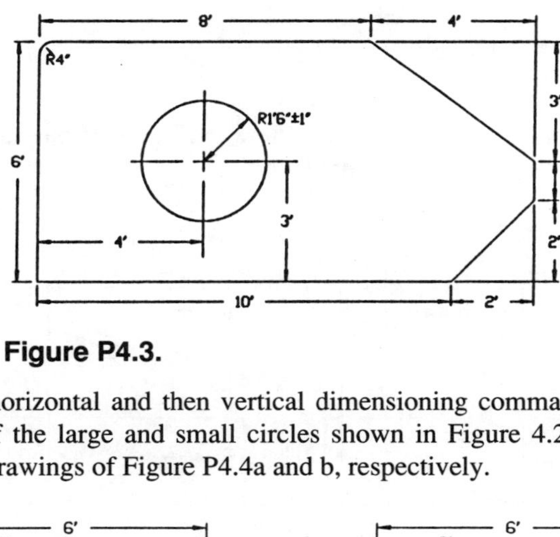

Figure P4.3.

4.4 Use first horizontal and then vertical dimensioning commands to detail the location of the large and small circles shown in Figure 4.2.4b to obtain the resulting drawings of Figure P4.4a and b, respectively.

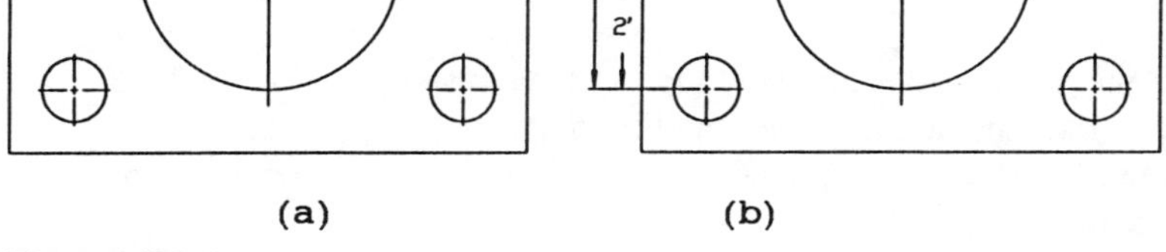

Figure P4.4.

4.5 Continue to add more dimensionings to Figure P4.4b by detailing the sizes of the plate first horizontally and then vertically as shown in Figure P4.5a and b, respectively.

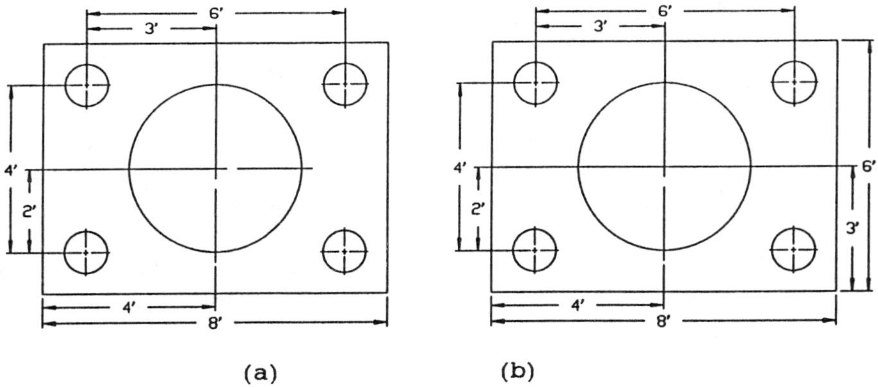

Figure P4.5.

4.6 Detail the radius of the small circle as shown in Figure P4.6a and then add the text line "4 Holes" as shown in Figure P4.6b.

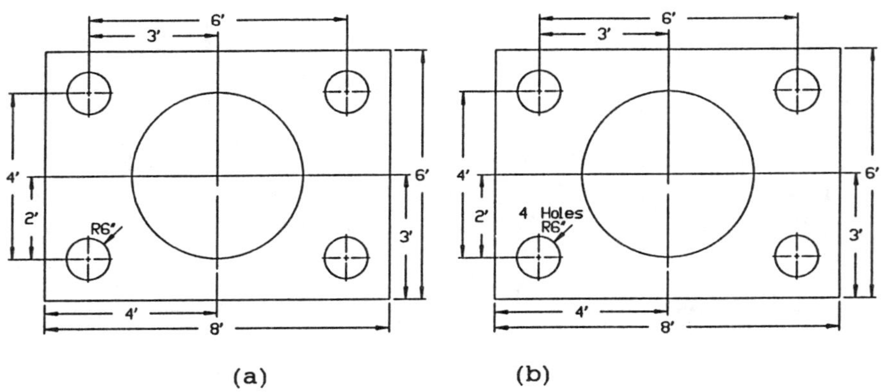

Figure P4.6.

4.7 Generate the points, curves, and captions as shown in Figure 2.7.5.
4.8 Generate the points, curves, and captions as shown in Figure 2.7.6.
4.9 Finish the dimensioning of the plate with five holes to have a final drawing as shown in Figure P4.9.

4.10 Add a border as shown in Figure P4.10 and then the title, name, and date to obtain a finished drawing of the holed plate.

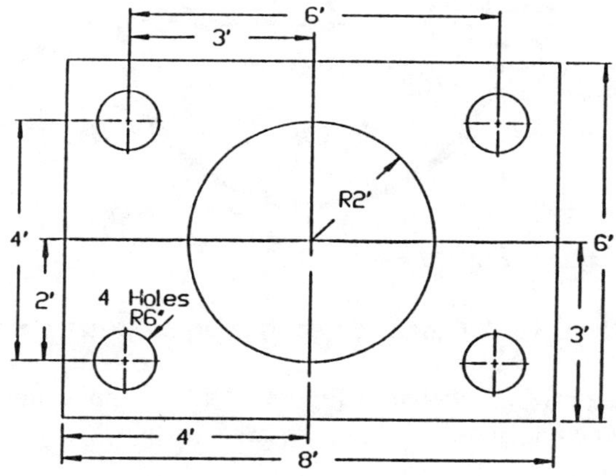

Figure P4.9.

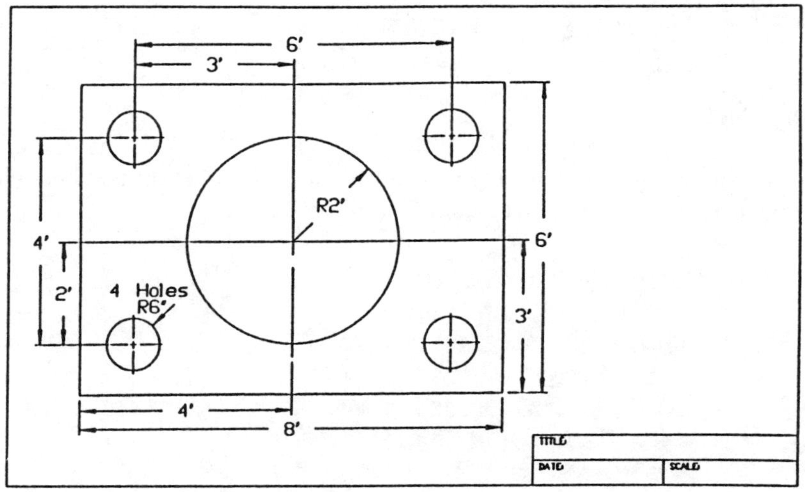

Figure P4.10.

4.11 Draw the wall clock shown in Figure P4.11 including the three needles, numeric texts, and radial tic marks.

Figure P4.11.

4.12 Add the text lines RESET and TEST onto the largest switch plate shown in Figure 2.12.

4.13 Recreate the drawings shown in Figure P4.13 by applications of appropriate **SilverScreen** commands.

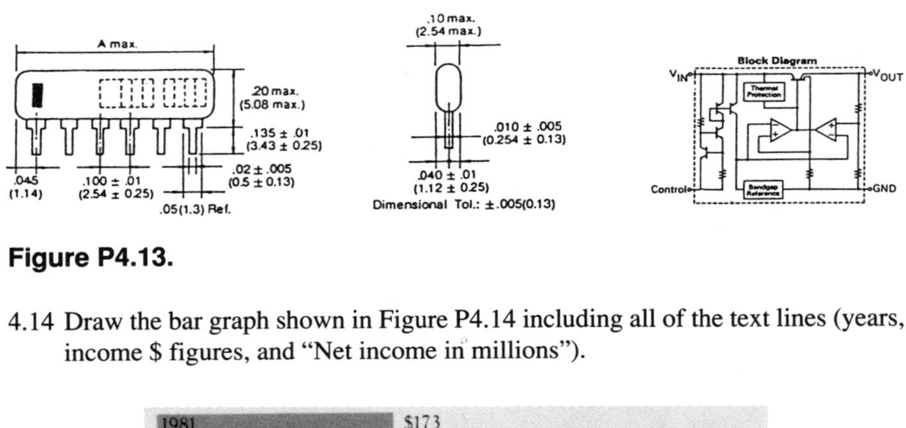

Figure P4.13.

4.14 Draw the bar graph shown in Figure P4.14 including all of the text lines (years, income $ figures, and "Net income in millions").

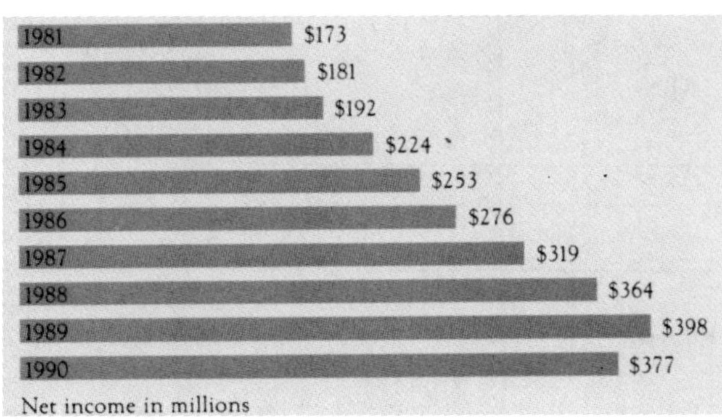

Figure P4.14.

4.15 Draw the "T" section shown in Figure P4.15 and then add the horizontal and vertical dimensions, and also add the text line "A T-Shaped Cross Section."

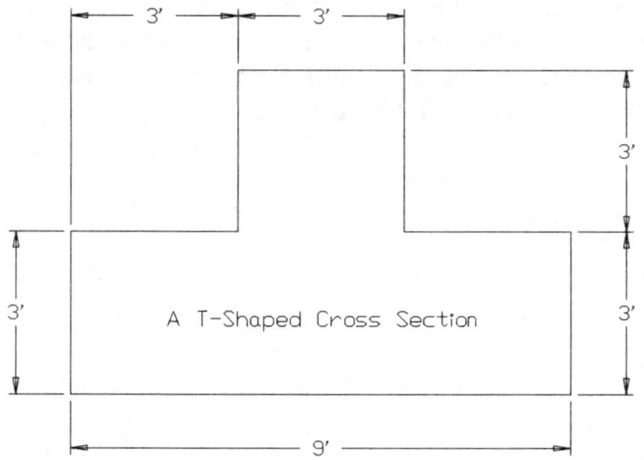

Figure P4.15.

4.16 Draw the unpolished version of a thin plate with three holes as shown in Figure P4.16. The corners have been filleted as indicated. Add all the detailings as marked.

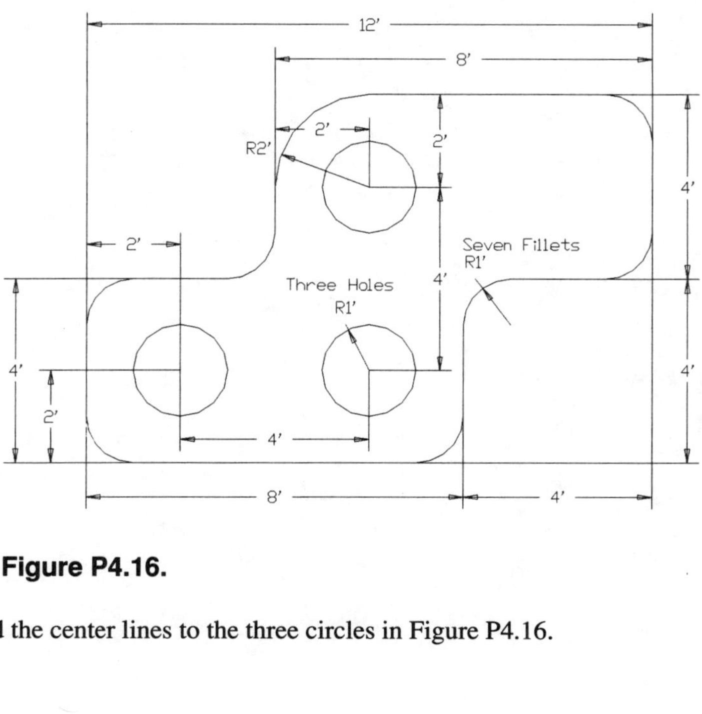

Figure P4.16.

4.17 Add the center lines to the three circles in Figure P4.16.

4.18 All fillets in Figure 4.16 have been using 20 linear segments for drawing a circle. These corners are to be refined by approximating the circular arc with more linear segments. Apply the **Refine** command of **SilverScreen** to do this.

4.19 Add a fourth circle also of radius 1′ at the upper right corner of the plate shown in Figure P4.16 to make the upper and lower halves of the plate look the same. After the addition, modify the detailing properly such as changing the "Three Holes R1′" to "Four Holes R1′" and so on.

5 Hard Copy, Storage and Retrieval of Drawing Files

5.1 INTRODUCTION

After a drawing has been satisfactorily generated, one certainly wants a permanent copy, either by printing or plotting it. Furthermore, if the drawing is worth keeping for future editing, it should be saved as a disk file so that redisplay of this drawing on screen can be expedited not by regenerating it through a step-by-step process, but simply by retrieving the saved **drawing file**. This chapter is devoted to a discussion of how to obtain a hard copy and how to save and retrieve **SilverScreen** files.

SilverScreen manages many different types of files. These files are distinguished by the **extension** associated with a **filename**. For example, Paof1234.DRW is a file having a name Paof1234 and an extension DRW. **DRW** indicates that it is a drawing file. Major extension names reserved for **SilverScreen** and extensively involved in this textbook are

BAS — files in **BASIC** language
C — files in **C** language
DRW — drawing files
DXF — data interchangeable files
EPS — postscript files
EX — executable files
FNT — font files
IGS — files in Initial Graphics Exchange Specification Format
PCX — Paintbrush files
PIX — snapshot files
PM — mesh-program files
SF — script files

We will discuss these files as the occasions arise. **SilverScreen** arranges to have the **EX** and **DRW** files (and others) stored in **EX** and **DRW** subdirectories of **Silver**, respectively. So, to look at all of the existing **DRW** files, for example, one first requests for a *c*hange of *d*irectory by having on screen (assuming that **Silver** is currently on the hard disk C)

C>cd DRW
C>dir (or, C>dir/p, or C>dir/w/p)

The C>dir (dir, abbreviated for the word directory, can be entered in either lower or upper case) asks for all files in the current directory or subdirectory to be listed one line per file on screen. If dir/p is entered, it will enable the listing to be stopped when the screen is full and the listing to be continued when the **Enter** key is pressed. This option circumvents having the listing of the files quickly displayed and passed on continuously to the end of files, which occurs when C>dir is entered. The dir/w/p option further shortens the listing of all files by arranging them in a matrix form so that several files can be listed in each line of the screen. This option also halts the listing process when the screen is full and resumes the listing when the **Enter** key is pressed.

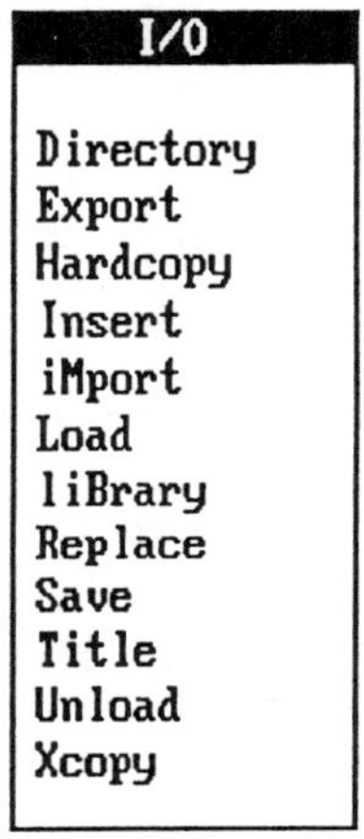

Figure 5.1.1.

The **Directory** command in the **I/O** menu of **Silver** (Figure 5.1.1), can also be employed for listing of all files in a subdirectory. For example, to list all files in the **DRW** subdirectory, one enters the command **Directory** to receive the prompt

Enter file specification:

We simply type C:\Silver\DRW*.* after the colon. To list all **SF** (script) files that are resided in the subdirectory **EX** of **Silver**, the entry should be

Enter file specification: C:\Silver\EX*.SF

5.2 HARD COPY OF SCREEN OR WINDOW DISPLAY

Most likely, what a user of **SilverScreen** would most like to have is a printed copy of whatever the user has already generated on the screen. To obtain a hard copy of a drawing being displayed on screen which may contain both graphs and texts, the **I/o** menu (Figure 5.1.1) of **SilverScreen** should be selected. The sequence of commands that needs to be entered, the screen prompts, and the responses for producing a hard copy shown in Figure 5.2.1a on a printer are

Hardcopy Print Screen Landscape Border Width
Enter Width (inches): 3
Enter number of copies: 1
Printer

One may choose to print either the drawings in the entire **Window Area** (see Figure 1.3.1) of the screen or the current window by entering either **Screen** or **Window**, respectively. The selected **Screen** or **Window** area can be printed horizontally or vertically by entering **Landscape** or **Portrait**, respectively. Figure 5.2.1b is the result when the command **Landscape** is changed to the command **Portrait**. Notice that the specified width (3 in.) is used by **Silver** as depth when the **Portrait** option is requested.

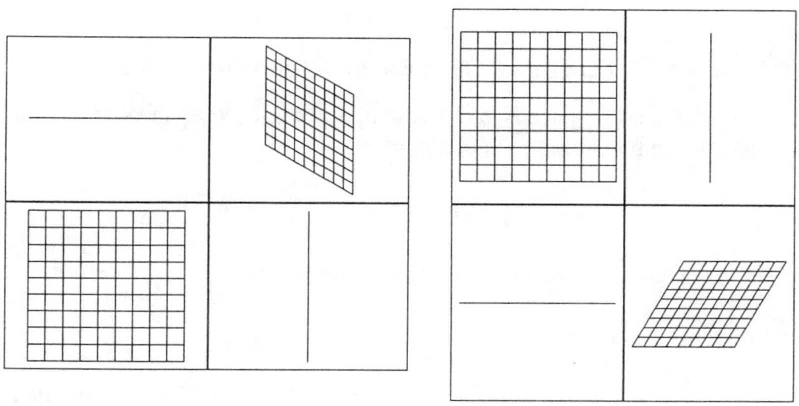

(a) (b)

Figure 5.2.1.

The **BORDER/NO-BORDER** option allows the display to be printed with or without a border. Figure 5.2.2 shows examples of the *current window* printouts: (1) using the **Portrait** option with a border and (2) after the isometric view being changed into the current window and then using **Landscape** and **NO-BORDER** options.

Plotters can also be used for making hard copies of what is displayed on the screen. In that case, a command **Plot** should be entered in place of the command **Print**. More involved commands would then be necessary for setting up the plotter. Readers should refer to the **SilverScreen** Reference Manual for further details. This text is primarily designed for beginners who have access to a simple printer.

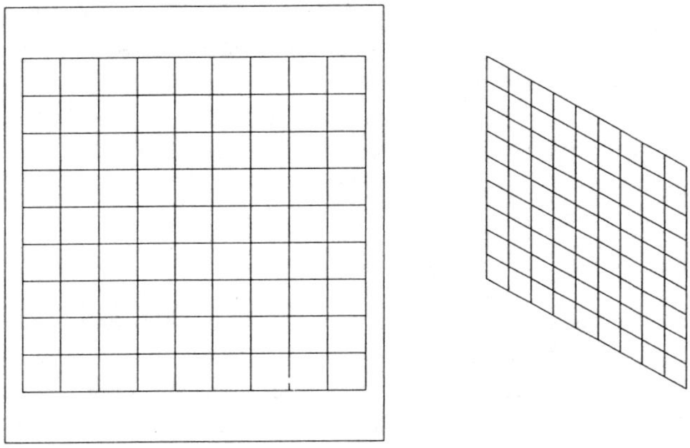

(a) (b)

Figure 5.2.2.

5.3 SAVING AND LOADING OF A SilverScreen DRAWING FILE

For future possible editing needs, a drawing being displayed on the screen should be saved as a disk file by entering the commands

I/o Save

At the top of screen, there will be a prompted question

Enter drawing name:

By providing a file name, say Temp, via keyboard input, and then pressing the **Enter** key, the displayed drawing is saved under that name. A new file named **Temp.DRW** will be created in the **EX** subdirectory of **SILVER**, which is the **root directory** of **SilverScreen**.

To retrieve a saved file, the commands are

I/o Load

Again, the user answers a prompted question, such as

Enter drawing name: Temp

Notice that there is no need to mention the extension DRW. Pressing the **Enter** key will result in redrawing of the saved image.

5.4 ORGANIZATION OF SilverScreen FILES

If a large number of files, drawings, and text lines have been created in application of the **SilverScreen** software, to organize them in an orderly manner becomes an important consideration for the purpose of quick retrieval of these files when the need arises. When a user begins to make a first drawing, **SilverScreen** automatically creates a block named **ROOT**, in which all drawings (primitives or objects) will be assmbled together, similar to the root of a tree on which trunk(s) and branches can grow. After an object has been created, an object name Obj1 will automatically be given unless the user specifies a desired name. The drawing, block, and object names are all displayed in the **Status Area** (Figure 1.3.1) in the form of

D:(none)	Drawing name
B:ROOT	Block name
O:obj1	Object name
3′6″	X coordinate of cursor
3′	Y coordinate of cursor
0′	Z coordinate of cursor
3″	current increment

If an object (obj1) is copied many times (by applying the **COPY** command in the **Geometry** menu, to be discussed later), indexed versions of new names Obj2, Obj3, and so on will automatically be assigned unless particular names are specified by the user. The user can rename block, drawing, object, and text line by applying the **Newname** command in the **Structure** menu (Figure 5.4.1). For example, to rename the object Obj1 with a new name Star, we enter the following menu, commands, and responses:

Structure Newname Object
Select (entity) to be named: Obj1
Enter new name: Star

New block can be created by applying the **Create** and **Block** commands in the **Structure** menu. And entities in a block can be relocated to another block using the **Relocate** command in the **Structure** menu.

As the number of entities continues to grow in the solid-modeling process of a complex structure or system, it is advisable to subdivide these entities into subblocks and/or groups. For example, Figure 5.4.2 shows that eight objects have been created. A stick is in the mug which together with a spherical ball, a washer, and a bolt are all on a tray. A cube and a pyramid are placed at the sides of the tray. We may therefore consider that there are three entities in the ROOT block, namely, Tray, Cube, and Pyramid. Tray can be considered as a subblock that contains Tray itself along with Ball, Washer, Bolt, and Mug. Mug is then a subblock of Tray and it contains Mug and also Stick.

STRUCTURE

Attribute
Create
Delete
Group
gOto
Join
List
Newname
Purge
Relocate
Tree
Undelete

Figure 5.4.1.

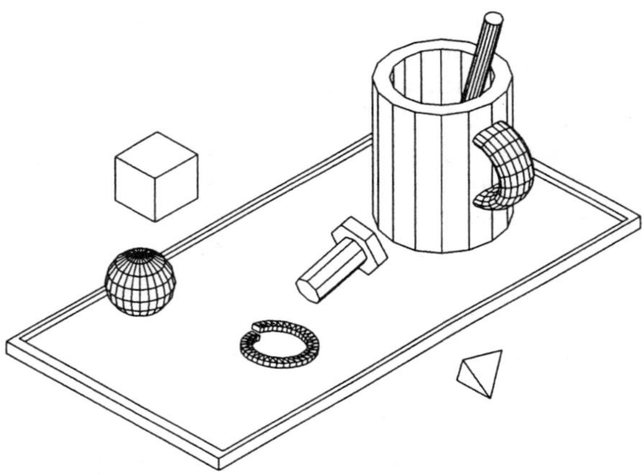

Figure 5.4.2.

Within a block or a subblock, one may sometimes have to select not all but only a few of the entities to form a temporary group for a certain modeling need such as to copy and rescale them. The **Group** command in the **Structure** menu can be applied to form a quick, **Q-group** for such a need.

In terms of file organization, a ROOT block is a ROOT directory and the subblocks are the subdirectories. The concept of directory and subdirectories is like a **tree** that has a trunk and then branches out to have many smaller branches. In the **SilverScreen** Reference Manual, the concept of such a tree structure is explained by using a **family tree** of Parents, Children, and Siblings.

RESTRUCTURING OF OBJECT FILES

Suppose that we have constructed eight objects (**tray**, **cube**, **pyramid**, **mug**, **ball**, **washer**, **bolt**, and **stick**) and they are all saved in the drawing file **Paof542.DRW** and in the ROOT directory. To obtain a printout of the tree structure of the ROOT directory, we can apply the **Tree** command in the **Structure** menu of **SilverScreen** by entering

Structure Tree Brief Print

The resulting printout is

BLOCK: ROOT [paof542] []
|— OBJECT: pyramid
|— OBJECT: stick
|— OBJECT: ball
|— OBJECT: cube
|— OBJECT: tray
|— OBJECT: mug
|— OBJECT: washer
|— OBJECT: bolt

If the objects in the tray, including **ball**, **washer**, **bolt**, **mug**, and **stick**, are to be grouped into a subdirectory to be named **tray**, the **Create** command in the **Structure** menu can be applied to create a BLOCK by entering

Structure Create Block At-root
Enter block name: **tray**

In case a created block needs to be deleted, the commands to be entered are

Structure Delete Block

The **F2** key should be pressed to pop-up all blocks on screen for selection of the one that is to be deleted.

To relocate the five objects, **ball**, **washer**, **bolt**, **mug**, and **stick**, into the newly created block **tray**, we first change from the ROOT directory to the subdirectory **tray** by entering the commands

Structure gOto Block

Again, the **F2** key can be pressed to pop-up all blocks on screen for selection of the desired one, in this case **tray**. When the **Enter** key is pressed, **B: tray** will appear in the **Status Area**. Now, we are ready to relocate the five objects one by one into the block **tray**. For example, to relocate the object **ball** we enter

Structure Relocate Object

The **F2** key is again pressed to pop-up all objects on screen for selection using a pointing device, or for responding to the question

Select object to be relocated: **ball**

In this manner, all five objects can be relocated into the block **tray**. A block named **mug** can also be created in the block **tray** and the object **stick** can be relocated into it to restructure the filing of the eight objects in **Paof542.DRW** in a Root–Child–Sibling way. The tree structure of this reorganized system can again be printed by using the commands **Structure**, **Tree**, **Brief**, and **Print**. The resulting printout is

```
BLOCK: ROOT [paof542] []
|— OBJECT: pyramid
|— OBJECT: cube
|— OBJECT: tray
|
+— BLOCK: tray
    |— OBJECT: bolt
    |— OBJECT: washer
    |— OBJECT: mug
    |— OBJECT: ball
    |
    +— BLOCK: mug
        +— OBJECT: stick
```

The advantage of organizing the constructed objects as a **tree** is that any single object or all objects in a named block can be isolated for presentation. For example, to display all objects in the block **tray** we enter

Presentation Isolate Block

The **F2** key can be pressed to have all blocks listed on screen for selection, and for responding the prompted question

Select block: **Tray**

Figure 5.4.3 is the split-screen, four-way display of the objects in the block **tray** when the objects tray, cube, and pyramid shown in Figure 5.4.2 are excluded.

To isolate any simple object, we simply request that object by using the commands **Presentation**, **Isolate**, and **Object**. Figure 5.4.4 is the resulting display of presenting the object **mug** alone.

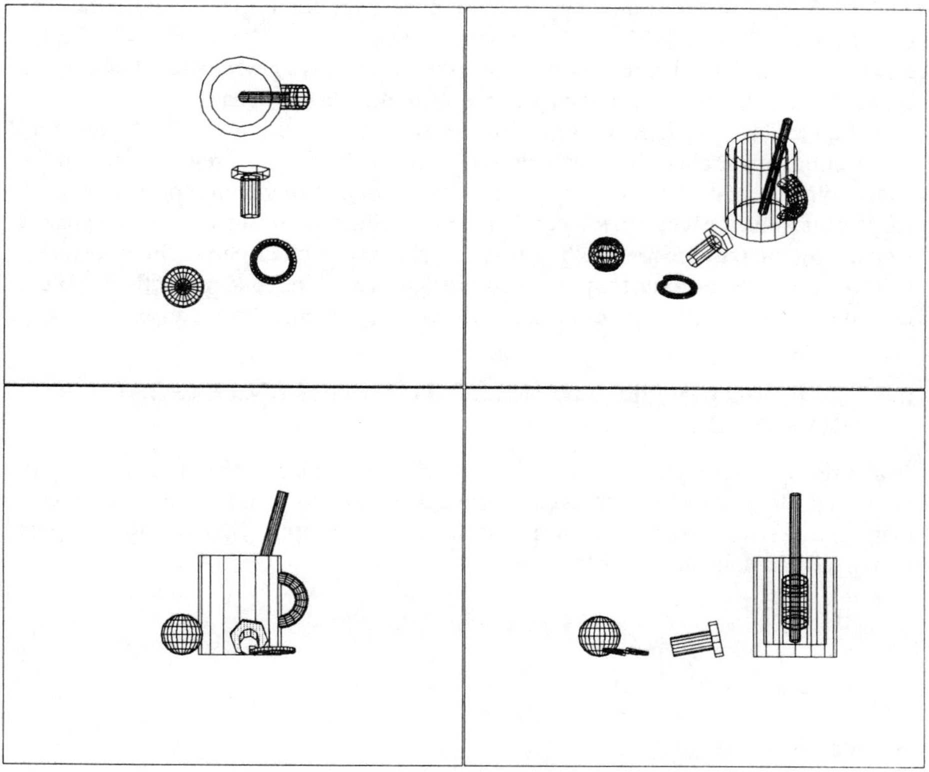

Figure 5.4.3. Five objects in the block tray.

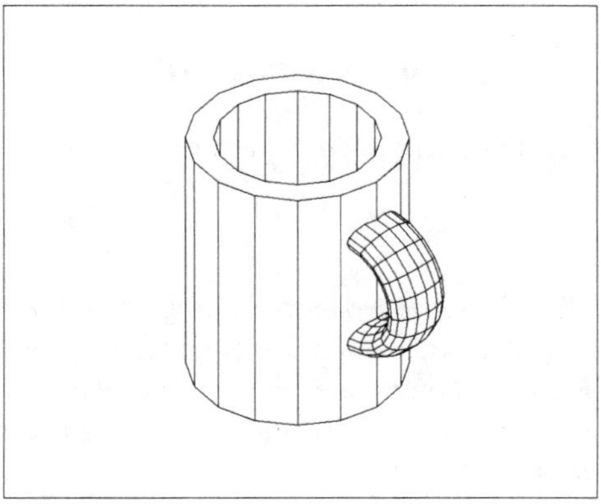

Figure 5.4.4.

To return to the ROOT directory, the commands **Structure**, **gOto**, and **Root** can be entered. It results in a display of **B: Root** in the **Status Area**. Another way to display all objects in the ROOT directory is to enter the commands **Presentation**, **Isolate**, and **Reset,** but this method will not change the **B:** in the **Status Area**.

In Figure 5.4.3, the object **tray** itself presently being included in the ROOT directory is not in the block **tray**. For completeness, it would be better to relocate it from the ROOT directory into the **Tray** block. Likewise, the object **mug** is presently in the block **tray** and it would be better to relocate it into the sibling block **mug**. Also, in the present setup, **mug** and **tray** are used both as an object name and a block name. One may prefer to have the blocks **mug** and **tray** renamed as **mugblk** and **Trayblk**, respectively. These considerations are left as homework problems for the readers to exercise.

5.5 DXF AND IGS FILES HANDLED BY IMPORT AND EXPORT COMMANDS

In addition to the commands **Save** and **Load**, **SilverScreen** has the option of saving or retrieving files that are in formats compatible with other software systems. A drawing displayed on screen can be saved in the Data Interchange Format (**DXF**) using an **Export** command. To do that, we select

I/o Export DXF

which results in a prompt of

Enter file name:

If the file name is Temp, an extension **.DXF** will be automatically added to make a complete file name of Temp.DXF.

To retrieve a **DXF** file, the commands should be

I/o Import DXF

and the user must respond to the prompted question

Enter file name:

If Temp.DXF is the file to be retrieved, either Temp or Temp.DXF could be entered via the keyboard because the extension .DXF is automatically assumed.

In addition to **DXF** files, **SilverScreen** can also retrieve files that are stored in the **IGES** (Initial Graphics Exchange Specification) format. The commands for retrieval of such formatted files are

I/o Import IGES

Notice that an **IGES** (not IGS) command should be entered.

5.6 SCRIPT FILES

The commands of **SilverScreen** used to generate a drawing of an object can be stored as a **script** file, which when later retrieved and executed will be able to redraw the object. For example, after the two-dimensional drawing shown in Figure 2.4.1 has been generated, we can request that the commands involved be saved in a script file **Paof241.SF**. This request should be made by entering the commands

Language Remember Begin
Enter script name: **Paof241**

Notice that the command **Remember** is in the **Language** menu (Figure 5.6.1) of **SilverScreen** and the extension **.SF** will be automatically added. This initiation of a script file must be arranged *before* any other **SilverScreen** command is entered. That

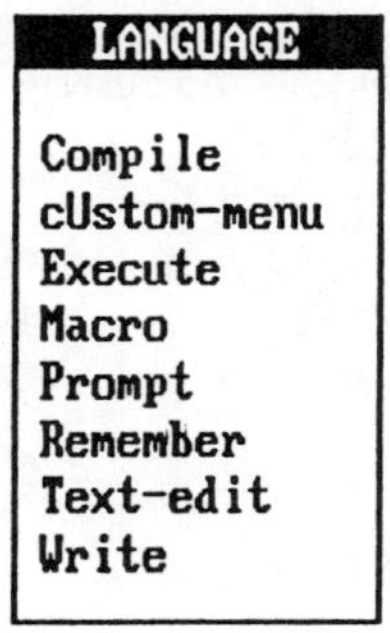

Figure 5.6.1.

is, after the above action of naming a new script file has been taken we can enter the other commands for the construction of Figure 2.4.1. According to Sections 2.4 through 2.6, the commands involved are

(Draw rectangle)
Draw Polygon Rectangle
Mark or enter corner of rectangle: 0,0
Mark or enter second corner of rectangle: 8,6
(Draw the small circle at the lower, left corner)
Draw Circle Radius
Mark or enter center of circle: 1,1
Mark or enter radius point: 1.5,1
(Draw the large circle)
Draw Circle Diameter
Mark or enter start of diameter: 4,1
Mark or enter second point of diameter: 4,5

(Draw the other three small circles)
Geometry Copy Primitive Cirlipse Matrix
Enter number of rows: 2
Mark or enter base point for displacement between rows:1,1
Mark or enter a second point: 1,5
Enter number of columns: 2
Mark or enter base point for displacement between column:1,1
Mark or enter a second point: 7,1

This script file **Paof241.SF** should be ended by reentering the commands **Language** and **Remember**. All script files are kept in the subdirectory **EX** of **SilverScreen**.

The content of this newly created script file **Paof241.SF** can be reviewed and displayed on screen by use of a DOS **type** command such as

C > type \Silver\Ex\Paof241.SF

The above statement assumes that C is the current disk drive. The following are the statements in the script file **Paof241.SF**:

```
draw rectangle 0,0,0 8,6,0
at 1,1,0
draw circle radius 0.5,0,0
at 4,1,0
draw circle diameter 0,4,0
copy primitive cirlipse \obj1.6 matrix
rows 2 delta y4 columns 2 delta x6
```

To retrieve and **continuously** execute to the end of the script file **Paof421.SF**, we enter the commands

Language Execute
Enter program or script name: Paof241
Continuous

The file can also be executed step by step at the control of the viewer who should press the **Enter** key to execute the next step. For such a request, the **Single-step** command should be selected instead of **Continuous** command.

5.7 CONCLUDING REMARKS

In this chapter, we have discussed the management of the files created in application of **SilverScreen** for wireframe and solid modeling. Files can be stored in many different forms so that they are compatible and can be accessible and used in conjunction with other computer software packages. For example, a drawing file stored with an extension **.PCX** can be read and then manipulated (such as rescaling, shearing, and rotating) by the popular word-processing software package WordPerfect.[1]

If there are changes that need to be made, files can be edited by applying the **Text-edit** command in the **Structure** menu of **SilverScreen**. If there are two script files that need to be appended together, the **Remember** and **Append** commands in the **Language** menu can be applied.

So far, only two-dimensional drawings have been created by use of **SilverScreen**. We are yet to generate three-dimensional solid models. A single, two-dimensional display such as the top view (Figure 5.7.1) of the assembly of a number of three-dimensional objects shown in Figure 5.4.2 simply cannot succinctly describe the true geometric characteristics of the objects. In the next chapter, we therefore commence the discussion of how three-dimensional objects are to be created and displayed by application of **SilverScreen**. When an existing object resided in a certain drawing file

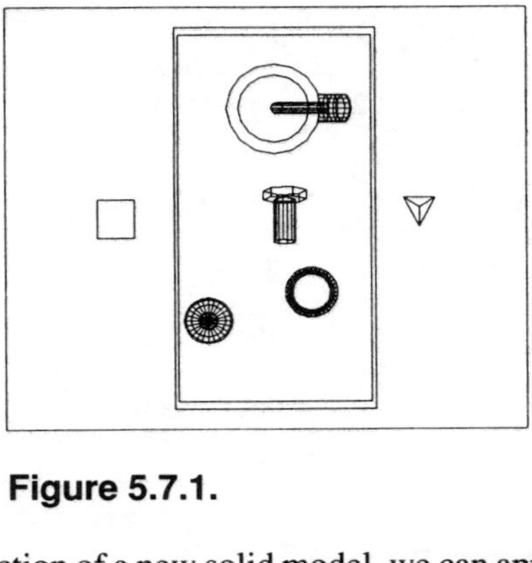

Figure 5.7.1.

is needed in the construction of a new solid model, we can apply an **Xcopy** method to extract it and append it into the current block. More on how the text line files for detailing on any selected plane using the **C-space** method should be managed will be presented in the ensuing chapters.

PROBLEMS

5.1 Create a script file **Paof231.SF** by application of the **Remember** command to save the commands listed in Section 2.3 for generating the illustration shown in Figure 2.3.1.

5.2 Run the script file **Paof231.SF** created in Problem 5.1 to display Figure 2.3.1.

5.3 Perform the exchange of windows so that the isometric view is placed in the current window and then produce a hard copy as shown in Figure 5.2.2b.

5.4 Repeat the drawing steps described in Section 4.2 to create a script file for Figure 4.2.2 and save it in **Paof422.SF**.

[1]Developed by WordPerfect Corporation, Orem, Utah 84057.

5.5 Run the script file **Paof422.SF** created in Problem 5.4 and then produce two hard copies using the **Landscape** option, one with border and the other without border.

5.6 Same as Problem 5.5 but using the **Portrait** option.

5.7 Append the script file **Paof422.SF** created in Problem 5.4 by detailing the drawing so that the result is as shown in Figure 4.3.1. Name this expanded script file **Paof431.SF**.

5.8 Produce a **Portrait** hard copy of Figure 4.3.1 by running the script file **Paof431.SF** created in Problem 5.7.

5.9 Load the script file **Paof431.SF** created in Problem 5.7 or any other script file onto **SilverScreen** and then apply the **Export** command in the **I/o** menu to save it as a paintbrush file with a name such as **Paof431.PCX**.

5.10 Apply the **Text-edit** command in the **Language** menu to change the script file **Paof241.SF** listed in Section 4.6 to remove the large circle so that Figure 2.4.1 is reduced to Figure P5.10.

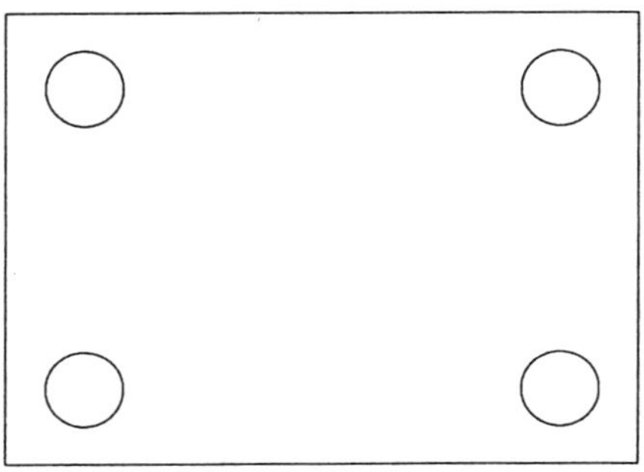

Figure P5.10.

5.11 In Section 5.4, the ROOT dierctory contains the objects **pyramid, cube**, and **tray**, and a block **tray**. Apply the **Newname** command in the **Structure** menu of **SilverScreen** to change the block name from **tray** to **trayblk**. In a similar way, change the block name **mug**. to **mugblk**.

5.12 After the name changes of the blocks have been done for Problem 5.11, relocate the object **tray** from the **ROOT** block to the **trayblk** block and relocate the object **mug** from the **trayblk** block to the **mugblk** block.

5.13 Apply the commands **Structure**, **Group**, and **Q-group** to form a quick group for the objects **stick, ball**, and **pyramid**, and then present them on screen by entering the commands **Presentation**, **Isolate**, **Q-group**, and **Current**.

6 Three-Dimensional Drawings and Objects

6.1 INTRODUCTION

When an object is thin such as a gasket, then its geometric shape can be described by connecting a series of points specified with two coordinate values. That is, these points can be treated as all on a Z = constant plane and only their X and Y coordinates need to be specified. We will introduce a number of commonly known three-dimensional curves and solids to lead the readers into the subject of three-dimensional engineering drafting and solid modeling by application of **SilverScreen**.

Spiral is a two-dimensional curve generated by a point continuously moving around and away from its initial location at constantly increasing distance. For a spiral, the locus of the moving point stays on the same plane, say $Z = 0$ plane. **Helix** is a three-dimensional spiral in which the locus is generated not only by increasing the radial distance from the initial position of that moving point but also by increasing the distance from the $Z = 0$ plane.

Pyramid is a three-dimensional solid with a polygonal base and triangular faces meeting at a common vertex. The simplest pyramid is a **tetrahedron**, in which the base is a triangle. Such a four-triangle-faced solid has six, straight-line connected edges. Let us denote the four vertices as points A, B, C, and D with three-dimensional coordinates (3,0,4), (3,0,–4), (–5,0,0), and (0,5,0), respectively. All units are in inches. Points A, B, and C are all on the $Y = 0$ plane and the point D is 5 in. above this plane. This tetrahedron can then be constructed by application of **SilverScreen** as follows, referring to Figure 6.1.1:

Draw Lines Multiple
Mark or enter start of first point: 3,0,4 (Point A)
Mark or enter endpoint: 3,0,-4 (Point B)
Mark or enter endpoint:
Enter endpoint: -5,0,0 (Point C)
Mark or enter endpoint: 3,0,4 (Point A)
Mark or enter endpoint: 0,5,0 (Point D)
Mark or enter endpoint: 3,0,-4 (Point B)
Draw Lines Single
Mark or enter start of line: 0,5,0 (Point D)

Mark or enter endpoint:
Enter endpoint: -5,0,0 (Point C)

It indicates that a series of connected lines, AB, BC, CA, AD, and DB, are first drawn by entering a multiple-line request. The tetrahedron is completed by following with a single-line request to connect the points D and C.

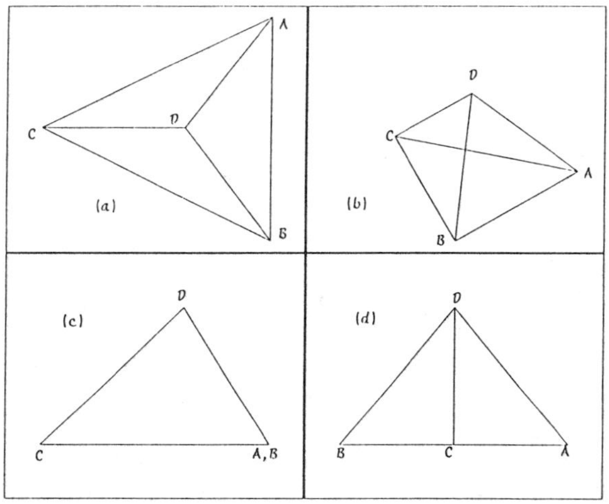

Figure 6.1.1. A four-way, split-screen display of a tetrahedron.

Since the constructed tetrahedron is a three-dimensional **wireframe**, there are various ways to display it on screen depending on how it is viewed or on the viewing angle. Figure 6.1.1 shows four different views of the constructed tetrahedron: (a) is viewed from the positive *Y*-axis direction, (b) is an **isometric** view, (c) is viewed from the positive *Z*-axis direction, and (d) is viewed from the positive *X*-axis direction. They are obtained by using **SilverScreen** commands

Window Split Four-way Drawing

It requests a four-way, split-screen for displaying the constructed tetrahedron.

Notice that the tetrahedron has a triangular base ABC and can be best viewed from the positive *Y*-axis direction as shown in Figure 6.1.1c, which is referred to as the **top view**. Figure 6.1.1c and d are referred to as the **front** and **right-side** views, respectively. In Figure 6.1.1c, vertices A and B coincide when they are viewed from the positive *Z*-axis direction.

Among the four different displays of the tetrahedron, Figure 6.1.1b gives a three-dimensional, **perspective** view while the other three appear more like planar views. In the following sections, we will introduce various commands of **SilverScreen** for construction of three-dimensional curves such as helix and commonly encountered three-dimensional solids.

6.2 GENERATING THREE-DIMENSIONAL CURVES USING THE F-MESH COMMAND

As in Chapter 2, the **F-mesh** command of **SilverScreen** can be effectively applied for generation of three-dimensional curves that are mathematically definable. As a first example, let us generate a circular cylindrical mesh by executing the following F-mesh program:

```
'Program CylindrM.PM — generates a circular, cylindrical
' spherical mesh of radius R, height H, and centered
' at (0,0,0)
'
    R  = 5
    H  = 10
   XC  = 0
   YC  = 0
   ZC  = 0
'
' Let A be the angle and E be the elevation.
'
    E: -H/2 to H/2 by H/10
    A: 0 to 2*PI in 18
    X  = XC+R*cos(A)
    Y  = YC+R*sin(A)
    Z  = ZC+E
```

This program **CylindrM.PM** should be stored in the **EX** subdirectory of **SilverScreen** so that the mesh can be easily displayed on screen by the commands and responses

Geometry f-mEsh
Enter name of file containing functional definition:
CylindrM
Select or enter horizontal color: 15
Select or enter vertical color: 15

The resulting display is shown in Figure 6.2.1.

As a second example, an f-mesh program called **Sphere.PM** to generate a spherical mesh is listed below.

```
' Program Sphere.PM — generates a spherical mesh of
' radius R and centered at (0,0,0)
'
```

R = 5
XC = 0
YC = 0
ZC = 0
'
' Let *A* be the azimuth and *C* be the colatitude.
'
C: –*PI*/2 to *PI*/2 by *PI*/18

A: 0 to 2*PI in 18
X = *XC*+*R**cos(C)*cos(A)
Y = *YC*+*R**cos(C)*sin(A)
Z = *ZC*+*R**sin(C)

The resulting display is shown in Fig. 6.2.2.

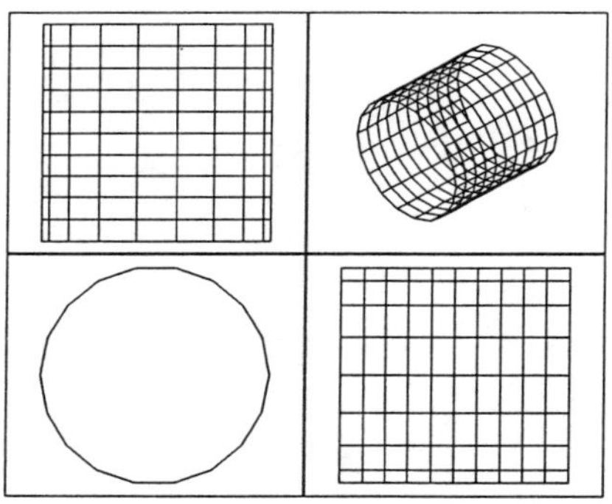

Figure 6.2.1. Cylindrical Mesh.

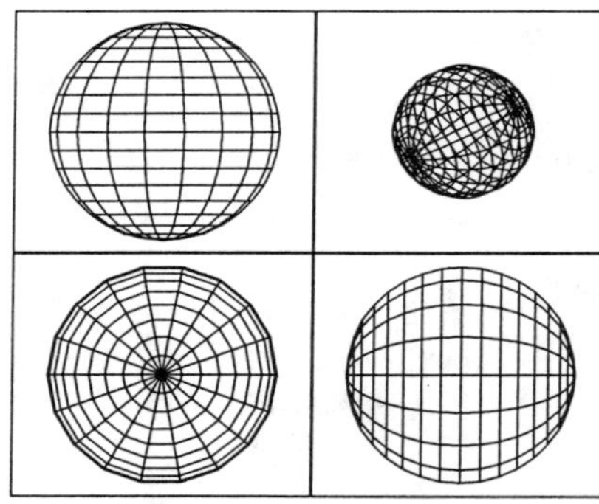

Figure 6.2.2. Spherical Mesh.

6.3 GENERATING A THREE-DIMENSIONAL SOLID BY LINEAR SWEEP

When a three-dimensional object has **uniform thickness**, then its geometry can be completely defined by giving the thickness and a detailed description of the shape and dimensions in the plane, which is perpendicular to the thickness direction. For example, the rectangle with five circles shown in Figure 2.4.1 could be the design of a prefabricated **brick**, which can now be completely defined if the thickness is specified.

SilverScreen has a number of **Sweep** options that can be applied to generate three-dimensional solids. The brick can be generated by a **Linear** sweep process. First, we load the file in which Figure 2.4.1 has been saved and have the **SilverScreen** menu displayed on screen. Suppose that the thickness of the brick is 1″ (all data for Figure 2.4.1 are in inches). The following commands result in a four-way display of the brick shown in Figure 6.3.1:

Geometry Sweep Linear Object
Press **Enter** key (Select obj1)
Mark or enter base point:
Enter base point: 0,0,0
(Select the left right corner as a base point.)
Enter target point: 0,0,1
(Base point and target point has a distance equal to the thickness of the brick.)
Enter scaling factor of sweep:1
(X, Y, and *Z* axes all have same scale factor. More examples are to be presented later.)

As a second example, Figure 6.3.2 shows that the three-dimensional solid can be generated by first drawing on the $Z = 0$ plane a rectangle having AB as one of its two diagonals and a circle having its center at O and a radius point at D, and then linearly sweeping it using either A or D as a base point and C or E as a target point, respectively. This drawing is ambiguous, in that one cannot distinguish whether the cylinder is a hole or a solid insert. This issue is to be resolved in a later chapter, which discusses Boolean operations of solids.

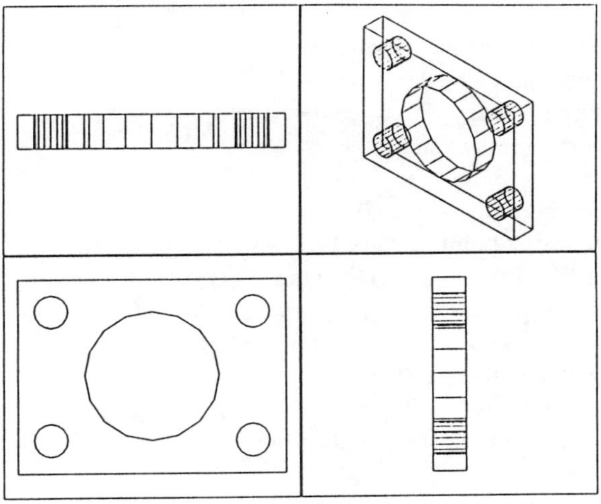

Figure 6.3.1. A solid generated by a linear sweep process.

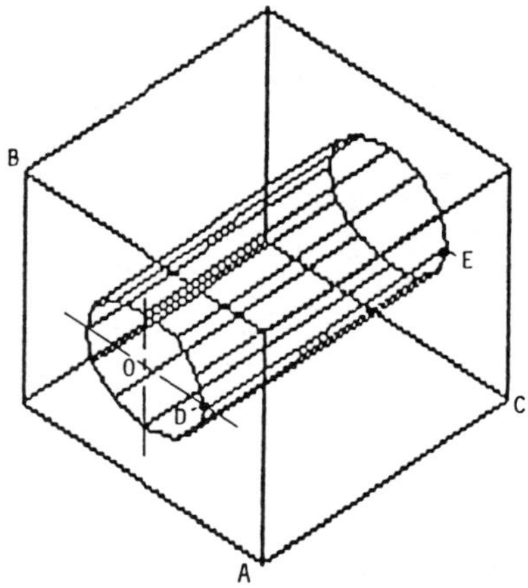

Figure 6.3.2. A linear sweep using the base point **A** and the target point **C**.

USES OF NONUNITY SCALE FACTOR

For linear sweep of primitives into solids, entering a scale factor other than 1 can sometimes be applied effectively for producing different solids. For example, we can make a solid of a "star" shape that does not have a uniform thickness but has a flat back surface by the entering following commands and responses:

Utility uNits Measure Inches
(request for units in inches)
Draw Polygon Star
Enter number of points in star: 5
Mark radius of star: 5
Mark or enter center of star: 0,0,0
Mark or enter vertex point: 2.25,0,0
Window Split Four-way Drawing
Geometry Sweep Linear Polygon
Mark or enter base point: 0,0,0 (center of star)
Mark or enter target point: 0,0,.5
Enter scale factor of sweep: 0
Enter name of object to be generated: **Star**

Based on the scale factor entered, 0, the resulting solid is a star having a flat back ($Z = 0$ plane) and a maximum thickness equal to 1/2 in. at its center and having the thickness linearly reduced to zero at all vertices. So the sweep process involves the formation of the lateral planes by connecting the target point with all of the vertices of the 5-point star. A four-way, split-screen display of this solid is shown in Figure 6.3.3.

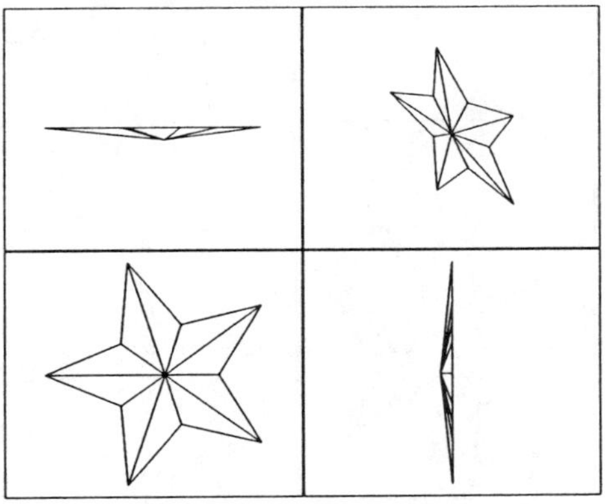

Figure 6.3.3.

A new solid, which is the mirror image of the generated solid, **Star**, relative to its back plane ($Z = 0$) can be easily constructed by the commands and responses to the prompted questions as follows:

Geometry Copy Object (enter **Star**)
mIrror Xy-plane
Enter or mark location of *z*-coordinate: 0,0,0
Enter name of object to be generated:
(By default, the name will be **Star1** if no name is entered.)

In Chapter 7, we will discuss methods by which solids of different shapes and sizes can be combined together and subtracted from one another, as well as other manipulations. In the present case, we are interested in knowing how the two flat back surface stars, with object names **Star** and **Star1**, can be united to form a true star!

Figure 6.3.3 is the result of using a scale factor equal to 0. If a scale factor of 0.5 were specified in the linear sweep of the star into a solid, we would have obtained a new solid as shown in Figure 6.3.4. It shows that the size of the primitive, the star-shaped polygon, is reduced to a half size and generated in the plane containing the target point and then the respective vertices on the two star-shaped polygons are connected to form the required lateral planes in the process of a three-dimensional solid. The case of scale factor equal to 0 is a special case when the size of the primitive is reduced to a **point** at the target point.

6.4 GENERATING AXISYMMETRIC SOLIDS BY CIRCULAR SWEEP

In Chapter 2, a polygon ABCDEFA was generated and shown in Figure 2.3.1. This closed contour can be used as a *generator* for creating an **axisymmetric solid** by rotating it about a specified axis. As shown in Figure 6.4.1a, if the axis of rotation is GH

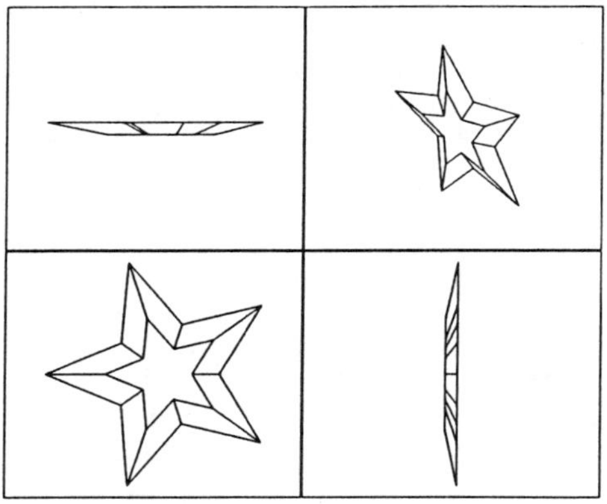

Figure 6.3.4. Linear sweep using a scale factor of 0.5.

we can generate the solid shown in Figure 6.4.1b by the following **SilverScreen** steps assuming that the contour ABCDEFA has already been defined:

Geometry Sweep Circular Object	
Enter top of axis: 0,1.5	(Point G)
Enter bottom of axis: 0,–0.25	(Point H)
Enter angle of sweep: 180	(1/2 revolution)
Enter number of faces: 15	(lateral polygons)
Select or enter horizontal color: 1	(blue)
Select or enter vertical color: 12	(red)
Enter name for object to be generated: **axisolid**	

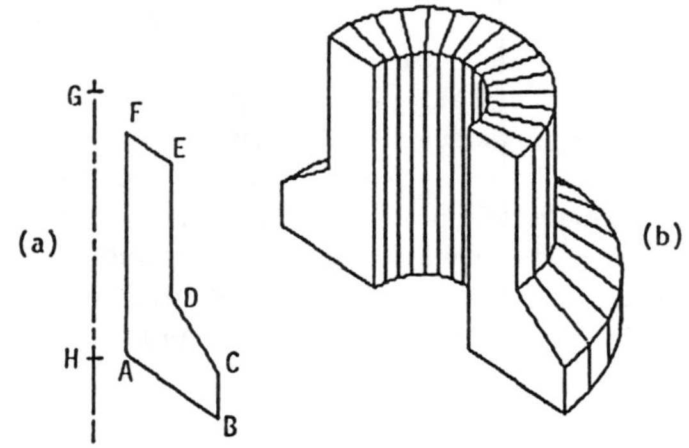

Figure 6.4.1. An axisymmetric solid shown in (b) is generated by a circular sweep of the polygon ABCDEFA about the axis GH shown in (a).

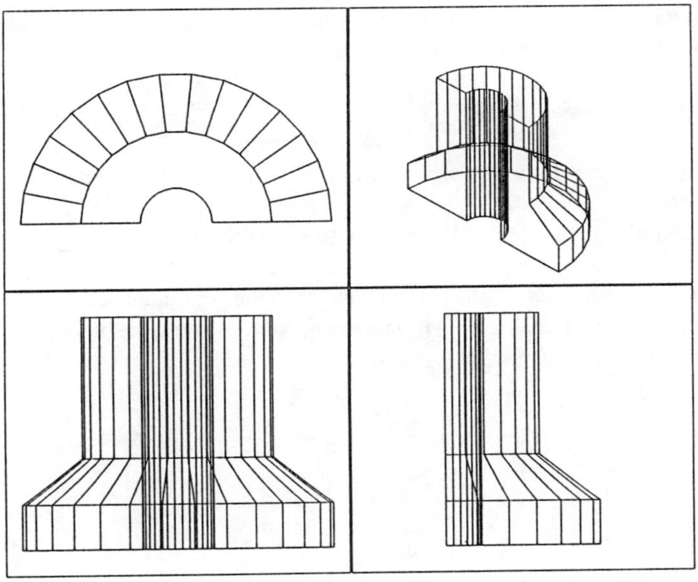

Figure 6.4.2. A four-way, split-screen display of the axisymmetric solid **axisolid**.

A four-way, split-screen display of this solid is shown in Figure 6.4.2. Invisible hidden lines have not been removed from all four views. For the isometric view, Figure 6.4.1b is the one for which the hidden-line removal technique has been applied. In Section 6.7, we will discuss the hidden-line removal capability of **SilverScreen**.

6.5 SWEEP USING THE JOIN COMMAND

Three-dimensional solids may also be generated by a sweep process of joining two primitives that are drawn on two different planes. For example, we may draw two squares of different sizes on two parallel planes and then join their respective vertices to form a three-dimensional framework, and then fill it to become a three-dimensional solid. Figure 6.5.1 is the result of applying the **Sweep** and **Join** commands in the **Geometry** menu using such a procedure. It first requires drawing two squares of size 8′×8′ and 1′×1′ on $Z=0'$ and $Z=-7'$ planes, respectively. The commands and responses to the prompted questions to be entered are

Draw Polygon Rectangle
Mark or enter first corner of rectangle: 8′,8′,0′
Mark or enter second corner of rectangle: 0′,0′,0′
Draw Polygon Rectangle
Mark or enter first corner of rectangle: 5′,5′,–7′
Mark or enter second corner of rectangle: 3′,3′,–7′

To join the two squares and then sweep into a solid, the commands and responses to the prompted questions to be entered are

Geometry Sweep Join
Select polygon:
(Use the pointing device or arrow keys to select the 8′×8′ square.)
Select polygon:
(Use the pointing device or arrow keys to select the 1′×1′ square.)

Figure 6.5.1a is a four-way, split-screen display of the generated solid whereas Figure 6.5.1b has the front and isometric window exchanged and then the hidden-line removal technique applied to the current window.

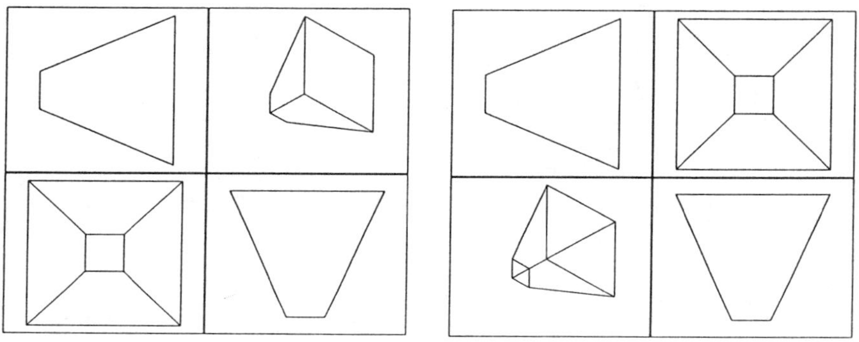

(a) (b)

Figure 6.5.1. A solid generated by a **Join** sweep, without filleting of its sharp corners.

This solid can be refined by rounding its sharp corners by application of the fillet capability of **SilverScreen**. However, the filleting must be done before the join process is applied. To do so, after the 8′×8′ square has been drawn we enter the following commands and responses to round the four corners:

Edit Fillet (Select the only object, 8′×8′ square.)
Insert All (Fillet all four corners.)
Enter radius of fillet: 1′

The next drawn 1′×1′ square on the $Z = -7'$ plane is to be filleted in a similar manner except a fillet having a radius equal to 0.125′ is used. The two rounded squares can now be joined together as before to generate a three-dimensional solid as shown in Figure 6.5.2.

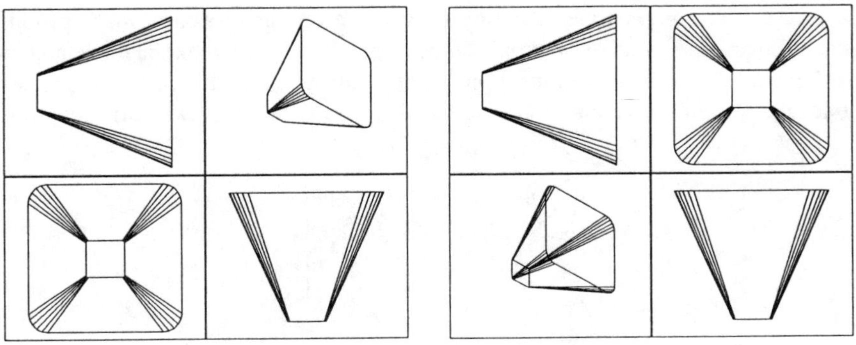

Figure 6.5.2. The same solid as in Figure 6.5.1 except the **fillet** command has been applied.

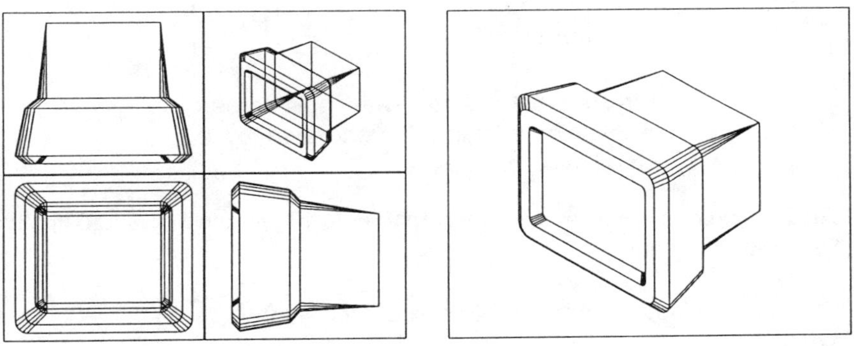

Figure 6.5.3. A solid model for the display monitor.

The procedure described above can be effectively applied to generate a solid model for the display monitor commonly used in computer systems such as the one shown in Figure 6.5.3. It is a stacked assembly of a number of solids having the basic form of that shown in Figure 6.5.2. Figure 6.5.3a gives three orthogonal views and an isometric view of the model whereas Figure 6.5.3b is the result of applying the hidden-line removal technique.

6.6 SPIRAL SWEEP

Figure 6.6.1 shows another feature of the sweep capability of **SilverScreen**. The steps of a stairway are assumed to have identical shape and dimensions. A step is first modeled as a **beam** of 25 units long constructed by a linear-sweep process using a rectangle of size 6 units deep and 12 units wide. An upright column with a height of

120 units can also be generated by a linear-sweep process using a circle having a radius equal to 30 units. A number of copies of the object (**beam**) can next be made and then positioned at different elevations along the outside surface of a vertically oriented, cylindrical column to simulate a stairway. In this case, the steps are **spirally** positioned.

Figure 6.6.1. A four-way, split-screen display of a stairway generated by a spiral-sweep process.

The **SilverScreen** commands and the responses to the prompted questions involved in this spiral sweep are

Geometry Copy Spiral Object (select the **beam**)
Angle
Enter angle of copy: 270
Enter number of copies: 15
Intercopy
Enter intercopy distance: 8
Mark or enter top of axis: 0,–40,60
Mark or enter bottom of axis: 0,–50,60

There are two options for specifying the range of the spiral, either in degrees or cycles. The former requires that an angle in degrees be specified such as 270° given above. Three-quarters of a revolution may also be specified using the cycle option. The number of copies entered (15 shown above) will divide the range of angle to arrive at an angle increment of (270/15 =) 18° for placing two neighboring steps at the intercopy distance specified (8 units shown above). The coordinates entered for the top and bottom points are on the axis of the vertical column, so it also serves as the axis of the desired spiral.

Figure 6.6.1 is a four-way, split-screen display of the spiral stairway. The upper, left window clearly indicates that the spiral does cover a range of 270°. Having all the lines that are supposed to be hidden behind the cylindrical column and thus invisible shown together with those visible ones, it is indeed difficult to conceive that the construct does represent a spiral stairway. In the next section, we discuss the procedure involved in obtaining a hidden-line-removed display. A display such as Figure 6.6.2 is needed to bring out the real view of the spiral stairway.

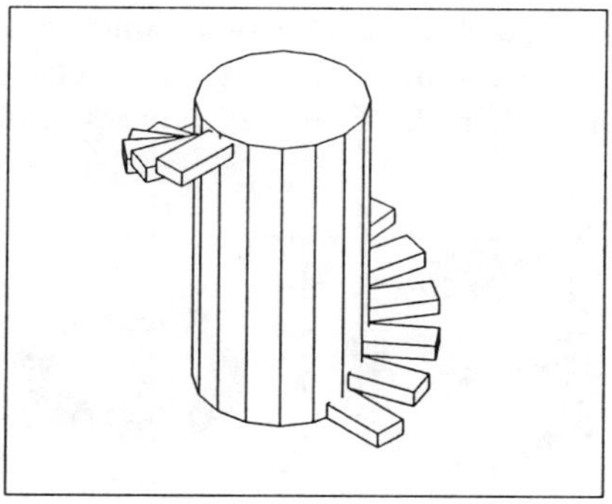

Figure 6.6.2. Isometric view of the stairway after removal of all hidden lines.

As another example of spiral sweep, let us generate a solid model of a spring washer, which is often used in industry for fastening purpose as a washer as well as a spring (Figure 6.6.3). First, we select the units in centimeters by entering the commands

Utility uNits Measure Centimeter

Next, we draw the generator of the spring washer, which is a rectangle by entering

Draw Polygon Rectangle
Mark or enter corner of rectangle: 5,5
Mark or enter second corner of rectangle: 6,5.5

For a spiral sweep of this rectangle into a spring washer, we enter the **SilverScreen** commands and the reponses to the prompted questions on screen as follows:

Geometry Sweep Spiral Polygon
Mark or enter top of axis: 8.5,5.5

Mark or enter bottom of axis: 8.5,5
Cycles
Enter number of cycles: 1
Enter intercycle distance: 0.75
Enter number of faces: 36
Enter name of object to be generated: **Washer**

The difference in the spiral sweeps between the stairway and the washer is that the former uses the **Angle** command and the latter uses the **Cycles** command. One cycle or one **revolution** is requested for generation of the washer. Also, as shown in Figure 6.6.4, which is a hidden-lines-removed isometric view of the spring washer, an angle increment of 10° is adopted in the sweeping process as a result of the 36-face specification.

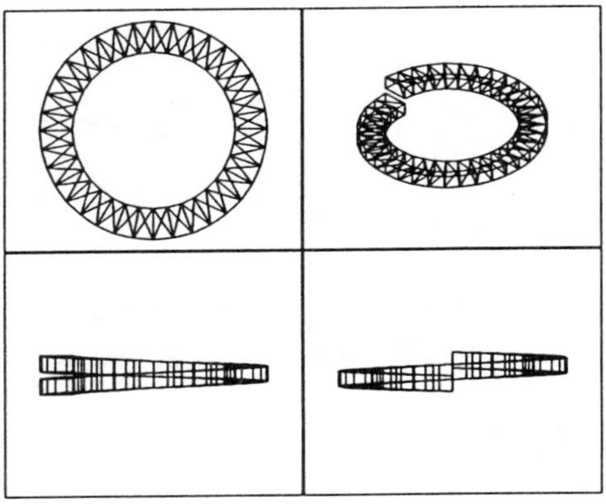

Figure 6.6.3. A four-way, split-screen display of the spring washer generated by a spiral sweep.

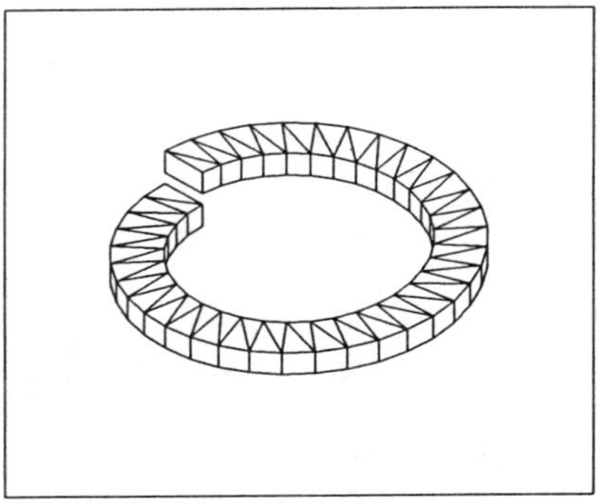

Figure 6.6.4. Isometric view of the spring washer after removal of all hidden lines.

6.7 HIDDEN-LINE REMOVAL

A solid model is three-dimensional and is often constructed using a large set of straight and curved lines. For example, the four-way, split-screen display of the half coupler shown in Figure 6.4.2 draws all lines that are needed to depict the object. Many of these lines are actually not visible and should be removed to present a better description of the object in a way such as the one shown in Figure 6.4.1b. In a more complex situation, one may have to apply a rather sophisticated scheme in order to produce a desired depiction of an object such as shown in Figure 6.7.1.

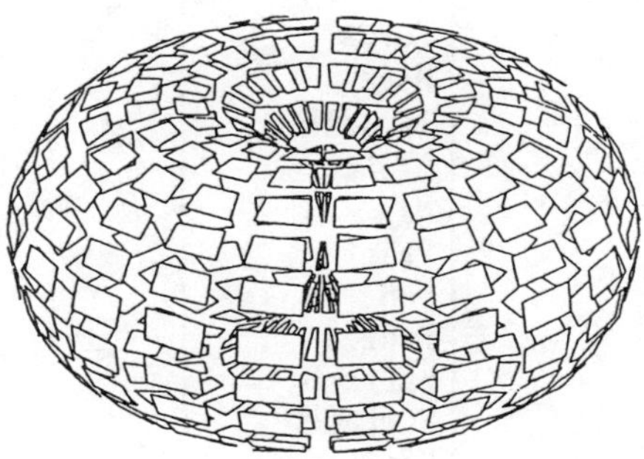

Figure 6.7.1 "Black Hole" drawn with a hidden-line removal technique. (Courtesy of **PDA Engineering**.)

To apply **SilverScreen** to display an object with all hidden lines removed, the object that is to be modified must first be placed in the **current window**. For example, to produce Figure 6.4.1b, we first exchange the isometric view at the upper-right corner of Figure 6.4.2 into the current window, which is at the left, lower corner. The **SilverScreen** procedure for doing that is to obtain a four-way, split-screen display by entering the commands

Window Split Four-Way Drawing

Next, we return to the main menu and select the commands

Window eXchange

By pressing the **Enter** key, the windows showing the isometric view and the front view are exchanged. If we desire to have an enlarged view, we return to the main menu and select the commands

Window Enlarge

By pressing **Enter**, the isometric view is enlarged.

To request removal of all hidden lines, we apply the **Hide** command in the **Presentation** menu (Figure 6.7.2). The sequence of commands to be entered is

Presentation Hide Drawing Hidden-Line

By pressing the **Enter** key, the solid is displayed with all hidden lines removed.

Figure 6.7.2. Presentation menu of **SilverScreen**.

Figure 6.7.3 is the result of applying the just described procedure of hidden-line removal to the five-hole plate previously shown in Figure 6.3.1. In this chapter, more hidden-line-removed drawings are shown in Figures 6.5.1b, 6.5.2, 6.5.3, and 6.6.2. In the ensuing chapters, this hidden-line-removed presentation will be extensively employed.

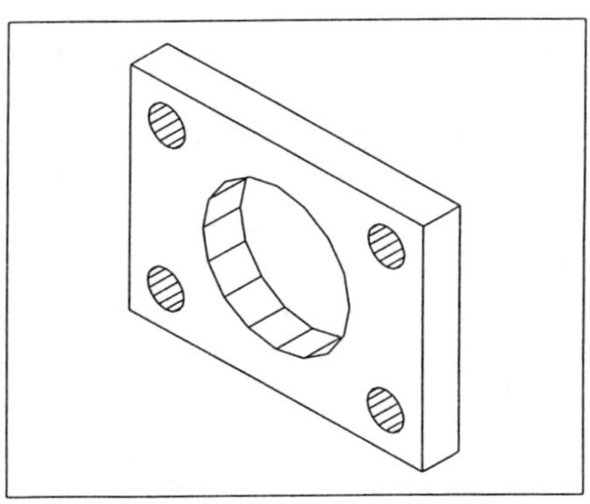

Figure 6.7.3. A hidden-line removed version of Figure 6.3.1.

6.8 CONCLUDING REMARKS

In this chapter, we have discussed various ways of applying the **SilverScreen** commands to construct three-dimensional wireframe and solid models, and also described how to present hidden-line-removed displays of such models. To distinguish whether an object is a wireframe or a solid, the hidden-line-removed presentation of the object can be effectively applied. For example, in Section 6.1, we have used multiple lines to construct a tetrahedron to represent a pyramid. It is a wireframe model and is not a solid because it can be seen *through*. If the hidden-line removal technique is applied to the window b shown in Figure 6.2.1, all lines will appear because the pyramid is only a skeleton and is not filled.

To construct a solid model for the pyramid by application of **SilverScreen**, we may use any one of the four sides of the tetrahedron as a primitive and the vertex, which is not in the plane of the selected side, as the target point in a linear-sweep process. Consider the case when the side ABC in Figure 6.2.1 is selected as the primitive and the vertex D as the target point. The **SilverScreen** commands needed for generating a pyramid by a linear sweep and the responses to the prompted questions are

Draw Polygon Triangle
Mark or enter first point of polygon: 3,0,4 (Point A)
Mark or enter endpoint: 3,0,–4 (Point B)
Mark or enter endpoint: –5,0,0 (Point C)
Geometry Sweep Linear Polygon
(Select the only triangle.)
Mark or enter base point:
(Use any one of the three vertices.)
Mark or enter target point: 0,5,0 (Point D)
Enter scale factor of sweep: 0
Enter name of object to be generated: **pyramid**

The resulting solid has a four-way, split-screen display as in Figure 6.2.1. Only when the hidden-line-removed presentation is requested, will this solid model be able to provide a display as presented in Figure 6.8.1.

In the next chapter, we will discuss how composite models can be formed by combining the solids that have already been constructed. Existing solids will also be modified, added, subtracted, and assembled by application of appropriate commands

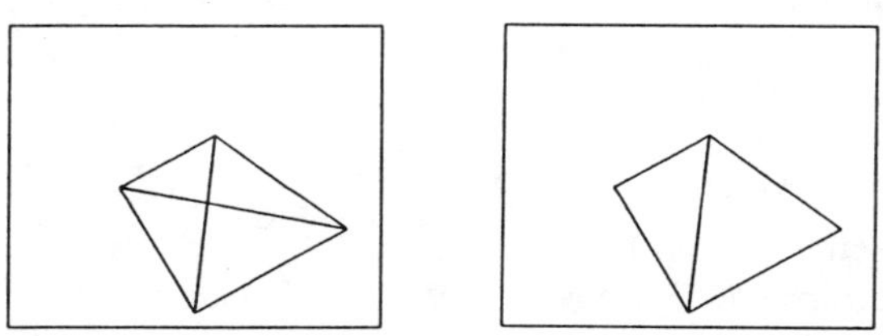

Figure 6.8.1. Wireframe Model (left panel) and Solid Model (right panel) of a tetrahedron.

of **SilverScreen.** It will be shown that wireframe models and solid models may also be combined. Erecting an antenna structure (approximated as a wireframe model) on the roof of a solid building is a good example. Use of the **Hide** command in the **Presentation** menu of **SilverScreen** will again be able to distinguish the wireframe components from the solid components.

Figures 6.3.1 and 6.3.2 are examples of showing a number of solid models simultaneously. But whether the circular cylindrical solids are insertions or holes remains to be clarified. The Boolean manipulations of solids to be discussed in the next chapter will help resolve this problem with the aid of hidden-line-removed presentation.

The solid model shown in the right panel of Figure 6.8.1 may appear as a quadrilateral polygon added with a line connecting two opposite vertices, instead of the representation of a pyramid as intended. We have to find a way to indicate that the two triangles actually represent two different surfaces of the four-sided pyramid. Toward that end, topics on shading the surfaces involved in a three-dimensional object by using different gray levels or colors, and changing the light source will be introduced in Chapter 9.

Before closing this chapter, let us construct a small collection of commonly used solids and store them in a drawing file called **PaoSolid.DRW**. Each of these basic solids can then be retrieved from this file, modified, and assembled together with other basic solids to form any desired combination. As a start, a tetrahedron, a hexahedron, a circular cylinder, and a sphere are included in this collection. The script file, **PaoSolid.SF**, to create these solids is listed below.

```
(Draw a tetrahedron)
at 0,0,0
draw equilateral vertex 3 delta 0,4,0
window split four-way drawing
sweep linear polygon \obj1.1 points 0,0,0 0,0,3.5 scale 0
name tetrahed
window split four-way drawing

(Draw a hexahedron)
at 10,0,0
draw equilateral vertex 4 delta 2,2,0
window split four-way drawing
sweep linear polygon \obj2.1 points 12,–2,0 12,–2,4.25
scale 1 name hexahedn

(Draw a circular cyclinder)
at 10,9,0
draw circle radius 2,0,0
display refine circle sweep 16
sweep linear cirlipse \obj3.1 points 12,9,0 12,9,7 scale 1
```

name **cylinder**
window split four-way drawing

(Draw a sphere — sweeping angle *incorrectly entered* as 360°
This is intentionally included to demonstrate the
consequence of such occurrence in building a script file.)
at 0,9,2.5
draw circle radius 2.5,0,0
display refine circle sweep 16
sweep circular cirlipse \obj4.1 axis 0,9.5,2.5 0,8.5,2.5
angle 360 faces 12 horizontal 15 vertical 15 name **sphere**
window split four-way drawing

(Draw a sphere — sweeping angle 180°)
at 0,8.5,2.5
draw circle radius 2.5,0,0
display refine circle sweep 16
sweep circular cirlipse \sphere.1 axis 0,9,2.5 0,8,2.5 angle
180 faces 12 horizontal 15 vertical 15 name **sphere**

A display of these four basic solids is presented in Figure 6.8.2. Other basic solids can be added onto this collection by applying the **Append** command of **SilverScreen** to modify an existing script file. Such an exercise is given in the problems at the end of this chapter for the readers to practice. In the next chapter, we will show how these basic solids can be retrieved from the file **PaoSolid.DRW**, modified, and combined with other solids to create some desired composite solids.

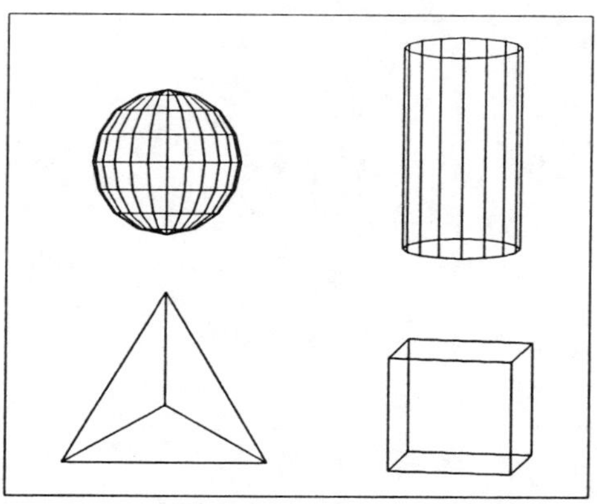

Figure 6.8.2. Some basic solid models constructed and saved in the drawing file **PaoSolid.DRW**.

PROBLEMS

6.1 Apply the **Copy** command of **SilverScreen** to the object, Half-Star, shown in Figure 6.3.3 for generating a *full* star as illustrated in Figure P6.1.

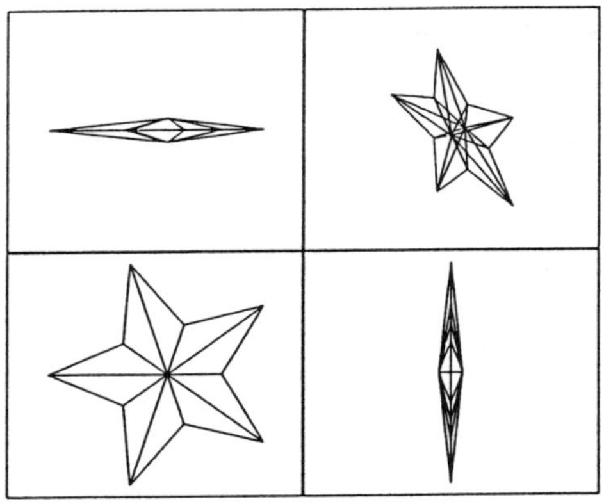

Figure P6.1.

6.2 Apply the **eXchange** and **Enlarge** commands in the **Window** menu of **SilverScreen** to place the isometric view shown in Figure 6.3.4 in the current window and enlarge it. Next, apply the **Hide** command in the **Presentation** menu of **SilverScreen** to obtain a hidden-line-removed display of the current window as illustrated in Figure P6.2.

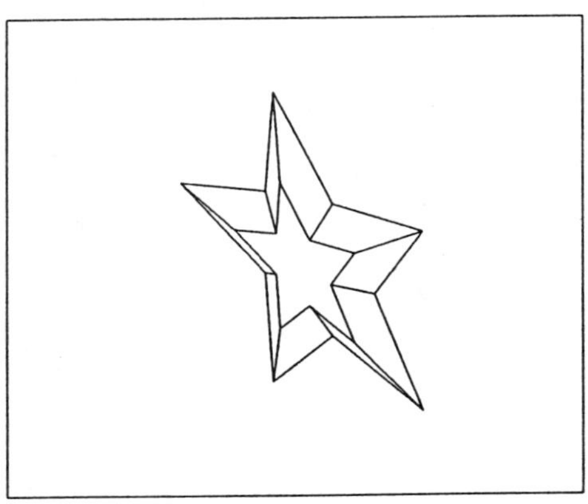

Figure P6.2.

6.3 Using a proper scale factor in application of the **Sweep** command in the **Geometry** menu of **SilverScreen** to generate the solid shown in Figure 6.5.3 by a linear-sweep process. The large 8′×8′ square on the Z = 0 plane is to be used as the primitive for sweeping and an appropriate point on the $Z = -7'$ plane is to be selected as a target point.

6.4 Construct the plate with a rectangular hole shown in Figure P6.4 without using the **Copy** command of **SilverScreen**. All four corners of the hole should be filleted with equal radius.

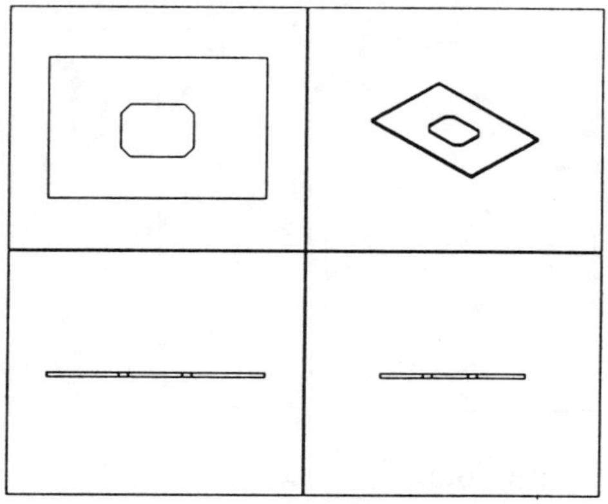

Figure P6.4.

6.5 Construct the plate with a circular hole shown in Figure P6.5 without using the **Copy** command of **SilverScreen**. The four corners of the plate should be symmetrically cut as indicated.

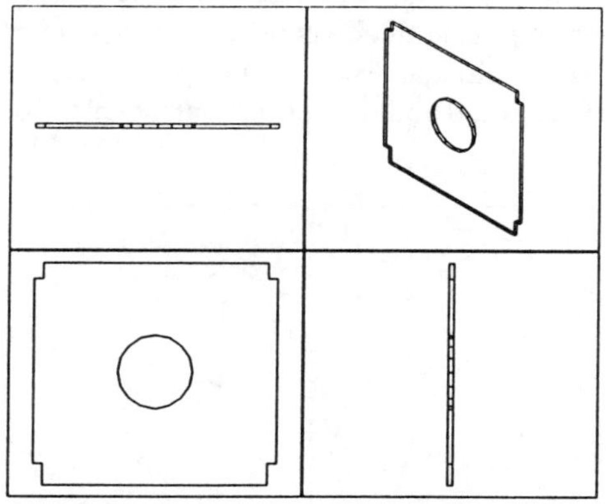

Figure P6.5.

6.6 Construct only the upper, left portion of the plate shown in Figure P6.4. The entire plate is to be completed by applying the **Copy** command of **SilverScreen**.

6.7 Construct only the upper, left portion of the plate shown in Figure P6.5. The entire plate is to be completed by applying the **Copy** command of **SilverScreen**.

6.8 Construct the plate with four identical, circular holes shown in Figure P6.8 without using the **Copy** command of **SilverScreen**. The plate also should have all of its four corners symmetrically cut.

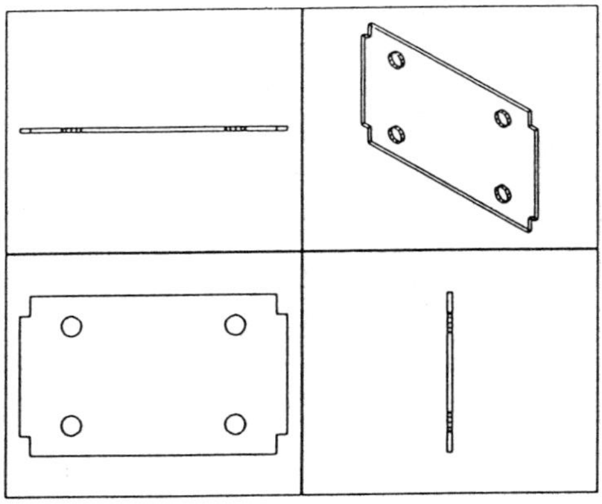

Figure P6.8.

6.9 Construct only the upper, left portion of the plate shown in Figure P6.8. The entire plate is to be completed by applying the **Copy** command of **SilverScreen**.

6.10 Create the script file **PaoSolid.SF** based on the description presented in Section 6.8.

6.11 Apply the commands **Language**, **Execute**, and **Continuous** to run the script file **PaoSolid.SF** for continuous generation of the four basic solids shown in Figure 6.8.2. Save the resulting display in the drawing file **PaoSolid.DRW**.

6.12 Apply the **Refine** command of **SilverScreen** to experiment with different approaches of modeling a circular disk to obtain approximated solids such as the two shown in Figure P6.12a and b having refinements equal to 10 and 20, respectively.

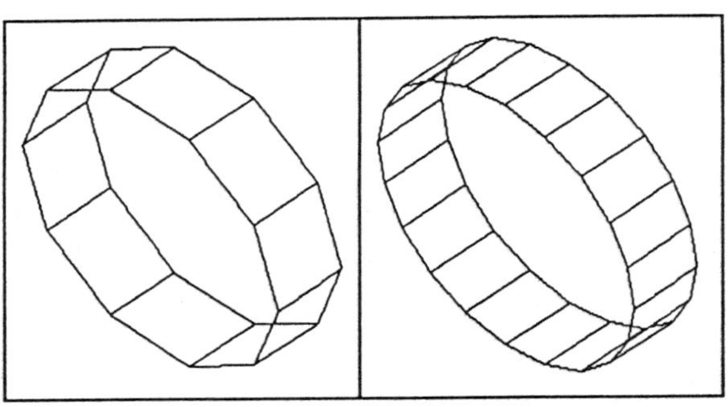

(a) (b)

Figure P6.12.

6.13 Construct solid models for the button-head machine bolt and oval-neck track bolt shown in Figure P6.13a and b, respectively.

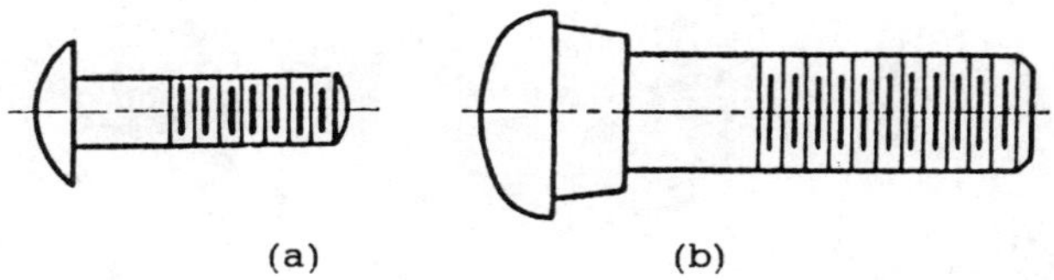

Figure P6.13.

6.14 The cube and cylinder shown in Figure 6.8.2 are obtained after appropriate rotations have been applied. Try the commands **A-rotate** and **Rotate** in the **Modify** menu of **SilverScreen** to obtain such a display on screen.

6.15 Construct a solid model for a shock absorber as shown in Figure P6.15.

Figure P6.15.

6.16 Construct a solid model for the shaft shown in Figure P6.16.

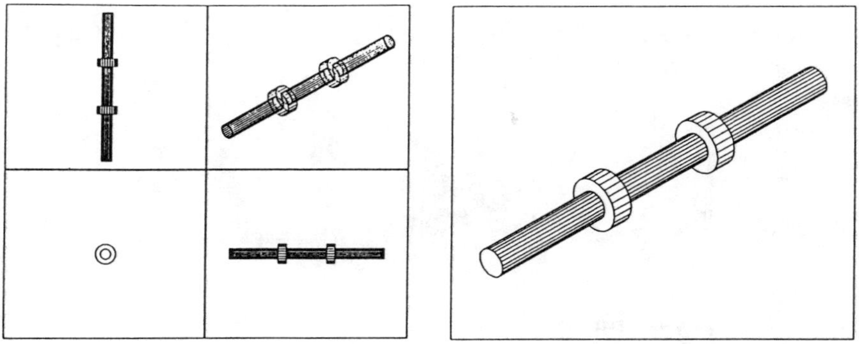

Figure P6.16.

6.17 Construct three-dimensional solid models for the stool, garden chair, and folding tray shown in Figure P6.17.

Figure P6.17.

6.18 Apply the **Sweep** command in the **Geometry** menu and the **Fillet** command in the **Edit** menu of **SilverScreen** to construct solid models for the prefabricated concrete blocks shown in Figure P6.18.

Figure P6.18.

6.19 Construct solid models for the storage hanger, electric light bulb, and conduit strap shown in Figure P6.19.

Figure P6.19.

6.20 Construct solid models for the switch plates shown in Figure P6.20.

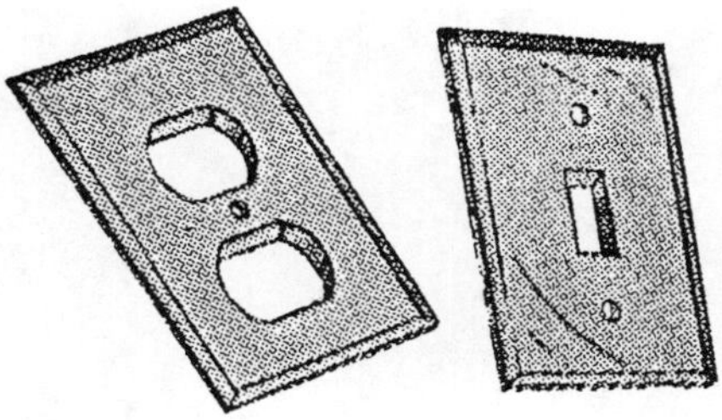

Figure P6.20.

6.21 Construct solid models for a standard key on a computer keyboard with a text on its top surface, Figure P6.21.

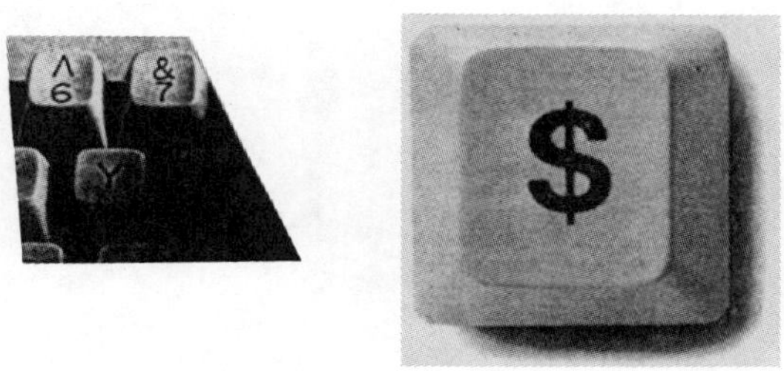

Figure P6.21.

6.22 Construct a solid model for a muffler as shown in Figure P6.22.

Figure P6.22.

7 Composite Solids

7.1 INTRODUCTION

Real-world objects can be modeled by combining various fundamental solids such as plates, spheres, prisms, and cylinders. Similar consideration can be given to the various shapes and sizes of possible **cutouts** in a solid. For example, the assembly shown in Figure 7.1.1 comprises two rectangular plates, a short hollow cone, and a short hollow cylinder, with two circular holes.

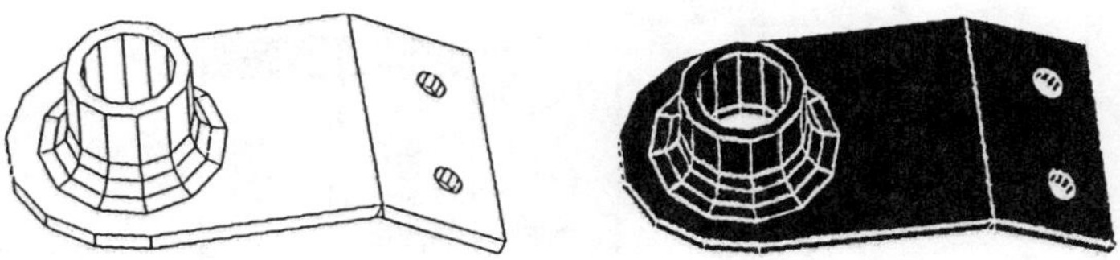

Figure 7.1.1.

SilverScreen has so far been shown capable of generating basic solids of simple shapes such as blocks, cylinders, cones, and spheres. When several solids have been simultaneously generated, their relationships will have to be clarified. For example, the block and cylinder shown in Figure 7.1.2, previously presented as Figure 6.3.2 in Chapter 6, overlap. The question "Does this block have a cylindrical hole or an insert?" will have to be answered. This chapter discusses the application of the **Boolean manipulation** capability of **SilverScreen**, such as subtraction and addition of two solids.

Also in this chapter, the **Modify** and **Geometry** menus (Figure 7.1.3) of the **SilverScreen** software will have extensive applications for construction of three-dimensional solids. The commands **Move**, **Rotate**, and **Scale** (enlarge or reduce) will be found most useful in translating, rotating, and varying the size and shape of a constructed solid before it is combined with other solids.

For example, the cylinder in Figure 7.1.2 could be moved halfway out of the front face of the cube and be subtracted from the cube to form a circular, cylindrical hole. The result is shown in Figure 7.1.4. Or the cylinder shown in Figure 7.1.2 could be reduced in size, copied four times, and then placed in a matrix form on the front face of the cube as shown in Figure 7.1.5. Use of the commands **Copy**, **Matrix**, **Solid**, **Boolean**, **Difference**, **Union**, and others in the **Geometry** menu of **SilverScreen** will be demonstrated repeatedly in the ensuing sections in this chapter.

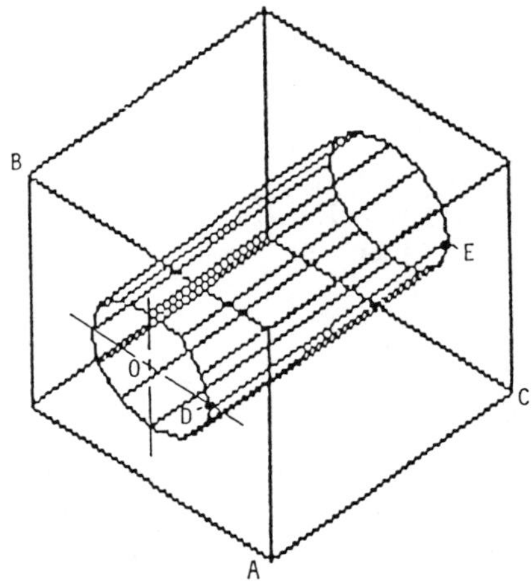

Figure 7.1.2.

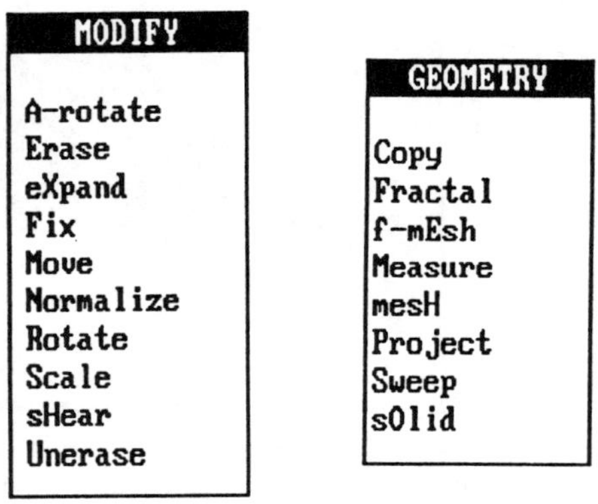

Figure 7.1.3. Modify and Geometry menus of SilverScreen.

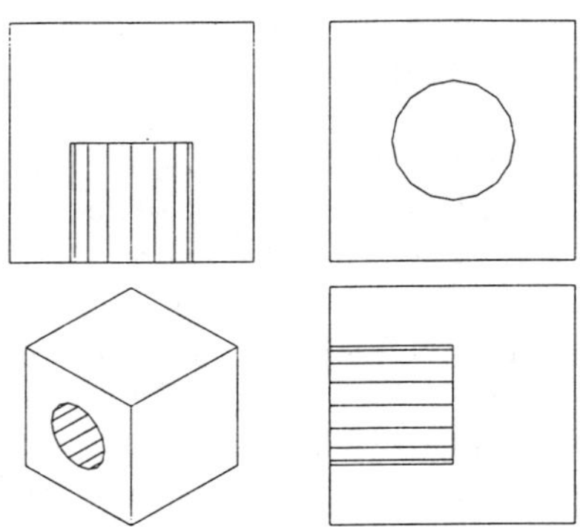

Figure 7.1.4.

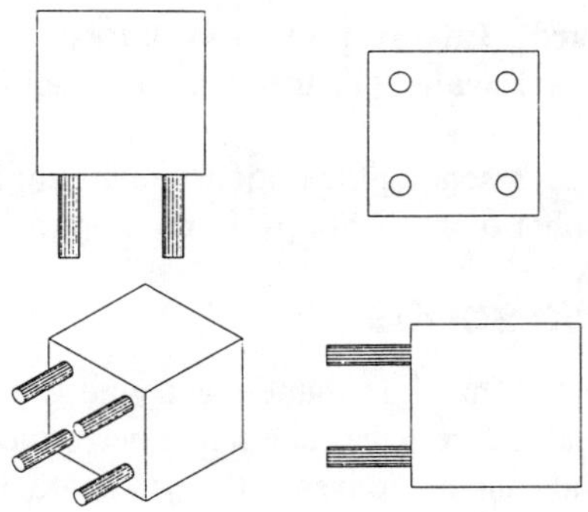

Figure 7.1.5.

7.2 BOOLEAN MANIPULATIONS OF SOLIDS

Let us first generate the cube and cylindrical rod shown in Figure 7.1.2 and then demonstrate how these solids can be manipulated to construct various composite solids. To construct the first object, the cube, we use the required **SilverScreen** commands and respond to the prompted questions as follows:

Draw Polygon Rectangle
Mark or enter corner of rectangle: 4,0,0 (Point A)
Mark or enter second corner of rectangle:
0,4,0 (Point B)
Geometry Sweep Linear Object
Mark or enter base point: 4,0,0 (Point A)
Mark or enter target point: 4,0,4 (Point C)
Enter scaling factor of sweep: 1
Enter name for object to be generated: **cube**

Likewise, the circular cylinder, named **rod**, can be constructed as the second object as follows:

Draw Circle Radius
Mark or enter center of circle: 2,2,0 (Point O)
Mark or enter radius point: 3,2,0 (Point D)
Geometry Sweep Linear Object
Mark or enter base point: 3,2,0 (Point D)
Mark or enter target point: 3,2,4 (Point E)
Enter scaling factor of sweep: 1
Enter name for object to be generated: **rod**

To combine these two solids (object names, **cube** and **rod**), the menu **Geometry** and commands **Solid** and **Boolean** can be applied. It results in the prompting of possible selections

Clip Difference Intersection iNvalidate interFerence
Slice seParate Trim Union Validate

Notice that the characters representing the options are highlighted in **uppercase**. We will illustrate below a number of applications of these options.

SUBTRACTION OF TWO SOLIDS

If the cube has a cylindrical hole having a shape exactly the same as the rod, a geometric subtraction of the cylindrical rod from the cube can be performed. This is done by using the **SilverScreen** commands and responses to the prompted questions as follows:

Geometry Solid Difference
Pick primary object: **cube**
(Press the **F1** key to get scan mode, maneuver mouse to move + cursor near the primary object, and press the left button of mouse.)
Selected object to be subtracted: **rod**
(Press the **F1** key to get scan mode, maneuver mouse to move + cursor near the object to be subtracted, and press the left button of mouse.)

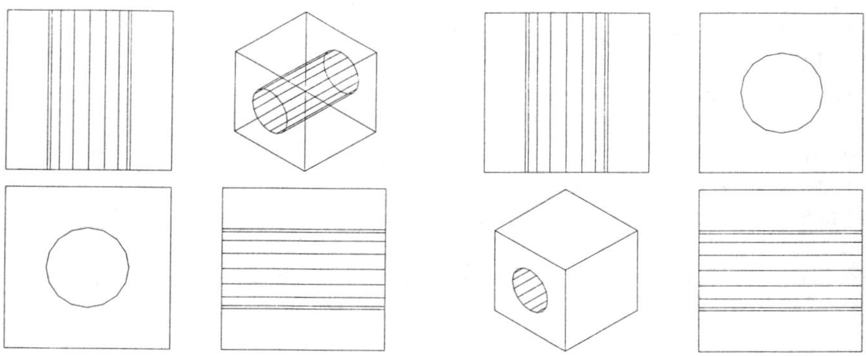

(a) (b)

Figure 7.2.1.

A four-way, split-screen display of the cube with a cylindrical hole is shown in Figure 7.2.1a. By exchanging the isometric view into the current window and removing the hidden lines, the display is changed to that shown in Figure 7.2.1b. It is still difficult to tell whether the cube has a hole or an insert! This ambiguity can be resolved when we add descriptive texts or shading to the drawing.

As another exercise of subtraction of two solids, let us move the rod halfway out of the cube along the positive *Z*-axis direction by application of the command **Move** in the **Modify** menu and then the commands **Solid**, **Boolean**, and **Difference** in the **Geometry** menu as follows:

> **Modify Move Object W-Space**
> Mark or enter base point for (W-space) move: 2,2,0
> (The center of the circular surface at the right end of the rod is selected.)
> Mark or enter second point: 2,2,2
> (The selected base point is moved 2 units along the positive *Z*-axis, consequently all points of the rod are so moved.)

The resulting, *moved* display of the two solids is shown in Figure 7.2.2.

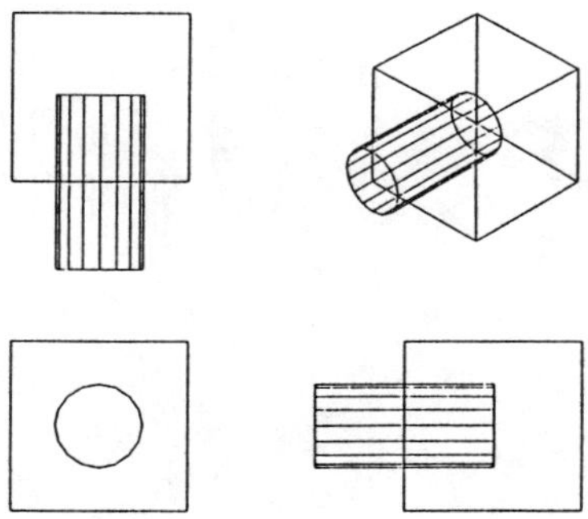

Figure 7.2.2.

After applying the **Solid**, **Boolean**, and **Difference** commands, the subtraction of the moved rod from the cube will produce a four-way, split-screen display as shown in Figure 7.2.3.

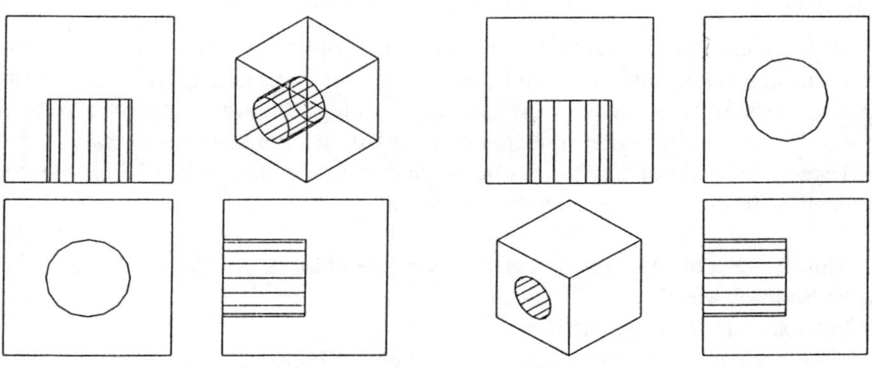

(a) (b)

Figure 7.2.3.

UNION OF TWO SOLIDS

Besides the possibility of performing the subtraction of one solid from another solid, solids can be added onto one another by application of the command **Union**. Again, we employ the two solids, **cube** and **rod**, shown in Figure 7.2.2 to demonstrate how the moved rod can be combined into the cube. The procedure is as follows:

Geometry Solid Union
Select first object: **cube**
Select second object: **rod**

It results in an entirely new object, for which a new name may be assigned. The resulting four-way, split-screen display is shown in Figure 7.2.4.

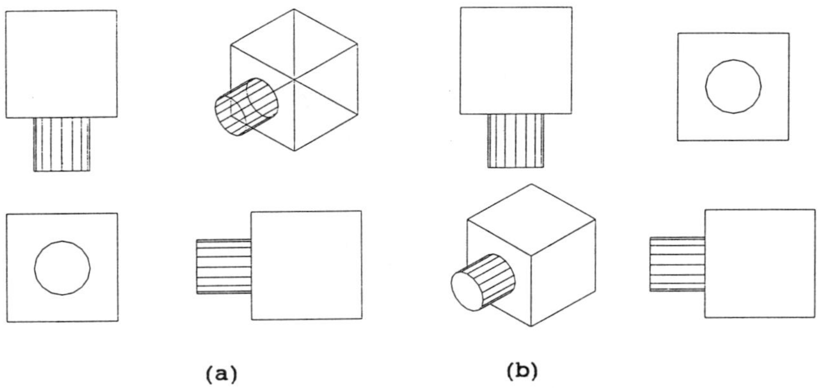

Figure 7.2.4.

7.3 COPYING OPERATION FOR REPETITIVE PROCESS

Suppose we need to construct a cube with four rods of the same shape attached to one of its surface as the cube shown in Figure 7.3.1. This is the extension of the "union" application of **SilverScreen** discussed in the preceding section except here the rods are of smaller size and *repeated* application of the **Union** command is necessary.

To reduce the size of the object **cube**, we request the **Modify** menu of **SilverScreen** for modification of the object and proceed as follows:

Modify Scale Object (selected the object **cube**)
W-Space Absolute
Enter scaling factor(s): 0.5,0.5,1
(The absolute *X*-, *Y*-, and *Z*-dimensions are reduced to

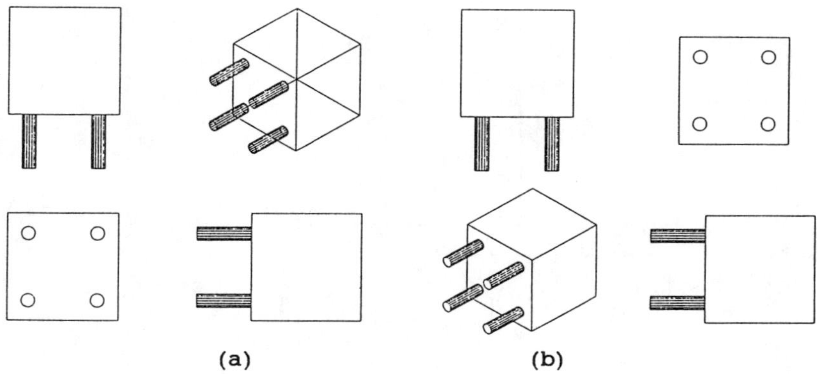

Figure 7.3.1.

0.5, 0.5, and 1 units from the original sizes of the rod of 2, 2, and 2 units, respectively.)

The next step is to move the reduced-size rod to one corner of the front surface of the cube and then copy it for the other three corners by the commands

Geometry Copy Matrix Object

Finally, the four rods must be unioned into the object **cube** to obtain the desired result as shown in Figure 7.3.1.

7.4 CONSTRUCTING A TYPICAL COMPOSITE SOLID — A DICE

In the last section, the **Geometry**, **Solid**, and **Boolean** commands are briefly explained for combining two basic solids to form a desired composite solid. Here, let us elaborate the procedure for construction of a dice. In the meantime, we will further demonstrate the application of the **Move**, **Scale**, **Rotate**, and **Copy** capabilities of **SilverScreen**.

The basic solids to be used are a cube and a circular disk. The size of the latter is to be changed, then moved to the six faces of the cube, and copied as many times as necessary to mark the faces of the cube (Figure 7.4.1). Boolean subtractions of a different number of circular disks from the cube are to be applied for all of its six faces.

CONSTITUENT SOLIDS

The cube selected is of size 4 by 4 by 4 generic units. It is generated by the commands **Polygon** and **Rectangle** in the **Draw** menu using the diagonal points of (0,0,0) and (4,4,0). This square is next linearly swept to obtain the desired cube by the commands **Sweep** and **Linear** in the **Geometry** menu with the base point (4,4,0) and target point (4,4,4). This object is then given a name, **cube**.

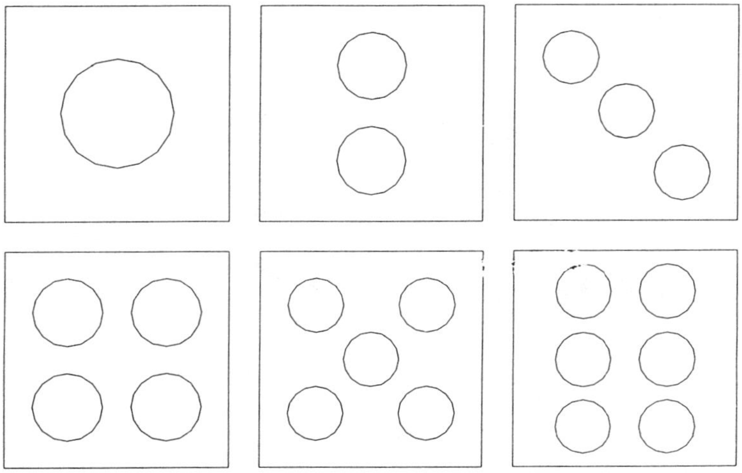

Figure 7.4.1. Six faces of a dice.

A circular disk is next generated by first drawing a circle in the current window (X–Y plane) using the commands **Circle** and **Radius** in the **Draw** menu. The center is at (2,2,4) and the radius point is (3,2,4). This circle is linearly swept into a disk by the commands **Sweep**, **Linear**, and **Cirlipse** in the **Geometry** menu using (3,2,4) as base point and (3,2,3.75) as target point. That is, the disk has a uniform thickness of 0.25 generic unit. This second object is given a name, **disk**. A four-way, split-screen display of the two objects, **cube** and **disk**, is presented in Figure 7.4.2.

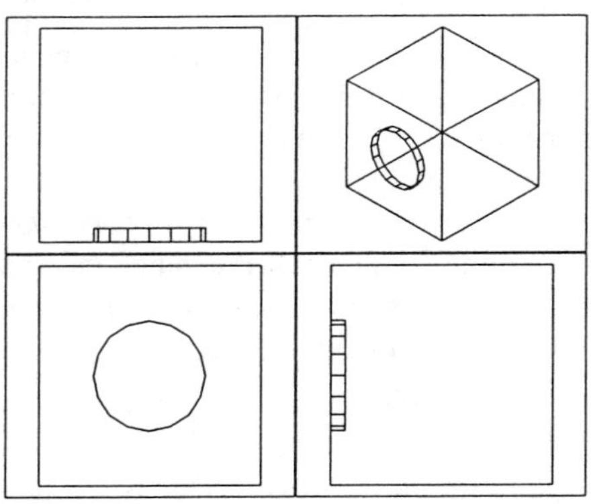

Figure 7.4.2. Two solids: cube and disk.

FACE WITH ONE DOT (Z = 4 PLANE)

Before the object **disk** is subtracted from the object **cube** to mark the *front* face of the cube as a first step of putting one dot on the dice, we copy it and create a third object named **disk1**. The commands used are **Geometry**, **Copy**, **Single**, and **Object**, and the selected object is **disk**. The new object, **disk1**, is placed at a distance away from the object **cube** using (2,2,4) as base point and (2,2,4.5) as target point.

Subtraction of the object **disk** from the object **cube** can now be implemented by using the commands **Geometry**, **Solid**, **Boolean**, and **Difference**. After the completion of the Boolean subtraction, a new object named **Cube1** is added to the **Silver** file directory and both files **cube** and **disk** are eliminated. In other words, the files were **Cube**, **disk**, and **disk1** before the subtraction and becomes **cube1** and **disk1** after the subtraction. Figure 7.4.3 shows the objects **cube1** and **disk1**.

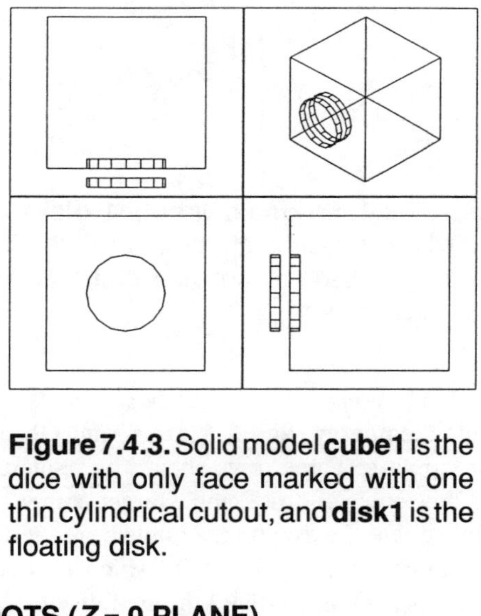

Figure 7.4.3. Solid model **cube1** is the dice with only face marked with one thin cylindrical cutout, and **disk1** is the floating disk.

FACE WITH SIX DOTS (Z = 0 PLANE)

To put six dots on the dice, we need to reduce the size of the disk, move it to the *back* face of the object **cube1**, copy it five times and place all six of them at the right locations, and finally subtract them from the object **cube1**. First, to *reducing* the size of the disk, we modify the object **disk1** by scaling the X and Y dimensions (both 2 units) by a half without changing the thickness (0.25 unit) in the Z-dimension. The commands used are **Modify**, **Scale**, and **Object**, and the scales entered should be 1, 1, and 0.25, which change the diameter of the disk from 2 units to 1 unit.

The reduced disk is next *copied* and *moved* to the *back* face of the object **cube1** by a *copying* process using the commands **Geometry**, **Copy**, **Single**, and **Object**. The point

(2,2,4.25), which is the centroid of the back face of **disk1**, can be selected as a base point and the point (2.75,0.75,0), which is the center of the lower, right circle (referring to the current, *X–Y* window), should be selected as a target point. The new disk is named **disk2**. Figure 7.4.4 shows a four-way, split-screen display of the objects **cube1**, **disk1**, and **disk2**.

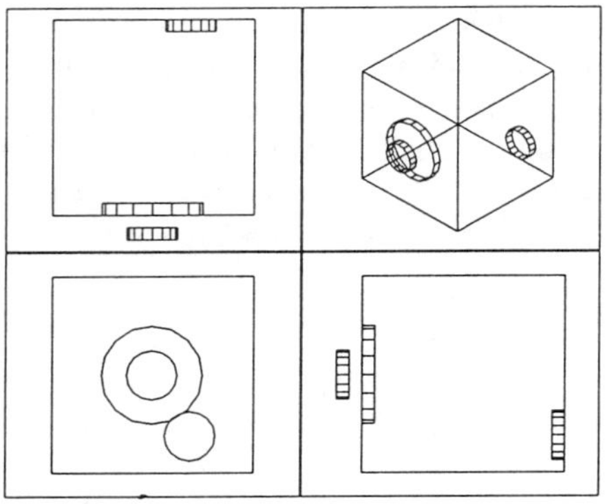

Figure 7.4.4. Solid models **cube1**, **disk1**, and **disk2**, which is on the *back* face of the dice (cube1) by copying **disk1** and then moved there.

Five more disks need to be constructed on the *back*, $Z = 0$ face of the object **cube1** by using the commands **Geometry**, **Solid**, **Matrix**, and **Object**. The object **disk2** should be selected for copying. Three rows should be requested with a base point (2.75,0.75,0), which is the center of disk2, and a target point (2.75,2,0), which is the center of the middle, right circle. Two columns should be requested with the same base point (2.75,0.75,0) and a target point (1.25,0.75,0), which is the center of the lower, left circle. These five new disks are named **disk3** through **disk7**.

Finally, the objects **disk2** through **disk7** are subtracted from the object **cube1** to create the transitory objects that have the names **cube2** through **cube7**, respectively. The final object **cube7** has one dot on the front ($Z = 4$) face and six dots on the back ($Z = 0$) face. At the end of six subtractions, the remaining files are **cube7** and **disk1**. Figure 7.4.5 shows a four-way, split-screen display of the objects **cube7** and **disk1**. **disk1** is being kept for the copying need of marking the other faces of **cube7**.

FACE WITH FIVE DOTS ($X = 4$ PLANE)

There are many ways to make five circular marks on the $X = 4$ face of the cube by application of **SilverScreen**. Here, we make use of the available floating object **disk1** and experiment with the **rotate** option in the **Modify** menu. Since the object **disk1** is oriented parallel to the face $Z = 4$ as shown in Figure 7.4.4, it needs to be rotated so that it will be parallel to the $X = 4$ plane and ready to be copied. The following steps accomplish the needed rotation:

Modify Rotate Object selecting disk1) W-space
Enter angle(s) of rotation: *Y*90

The requested rotation of 90° about the *Y*-axis resulted in changing Figure 7.4.5 into Figure 7.4.6.

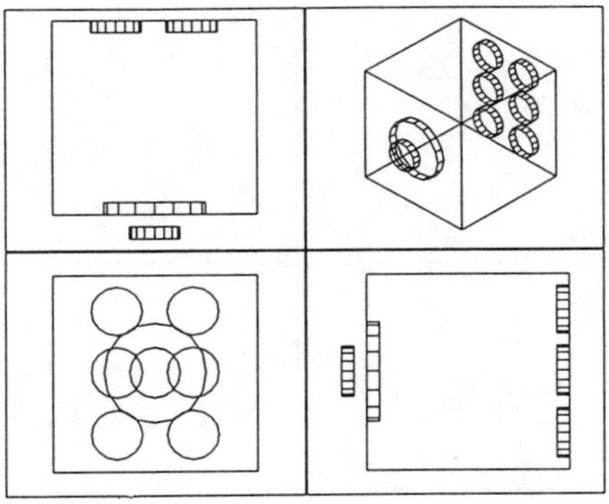

Figure 7.4.5. Solid models **cube7**, which is the dice having its front and back faces marked with one and six thin cylindrical cutouts, respectively, and **disk1**, which is needed for marking the other faces of the dice by the copying process.

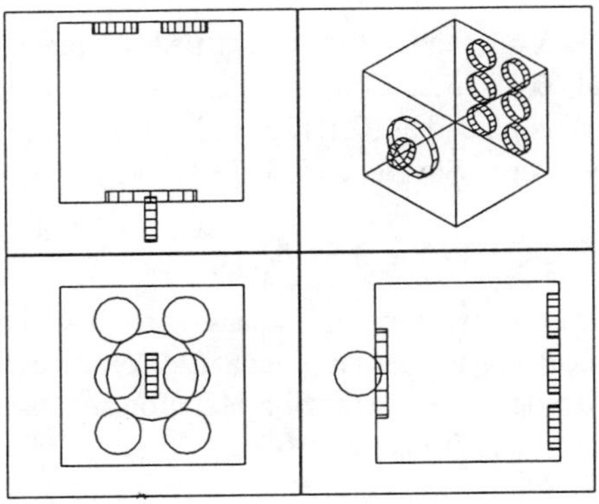

Figure 7.4.6. The floating object **disk1** has been rotated 90° about the vertical, *Y*-axis.

Next, we follow the procedure, previously used for the face with six dots, to copy the rotated disk five times and place them at proper locations on the $X = 4$ face of the cube. Figure 7.4.7 is the result of (1) copying the rotated disk, **disk1**, and placing the center of one copied disk at (4,2,2) and another copied disk at (4,3,1) by use of the commands **Geometry**, **Copy**, **Single**, and **Object**, and (2) copying the newly placed, upper right disk using the **Matrix** option and by a request for two rows and two columns which results in having five disks properly placed on the $X = 4$ face, and (3) subtracting these five disks from the cube.

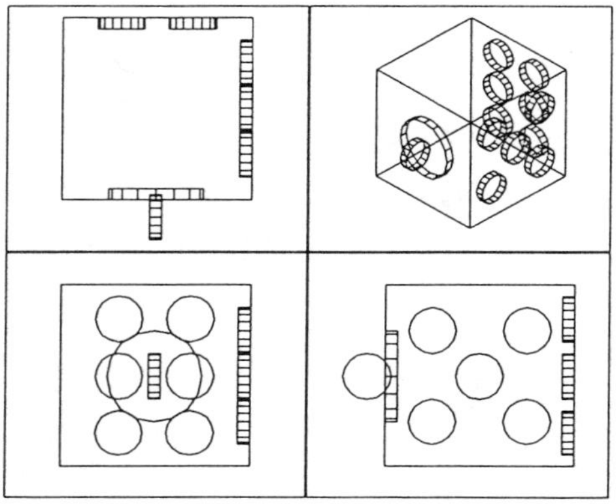

Figure 7.4.7. A dice has three of its six faces marked, one dot on the $Z= 4$ face, six dots on the $Z=0$ face, and five dots on the $X= 4$ face. The floating disk, **disk1**, remains for more copying needs.

FACE WITH TWO DOTS ($X = 0$ PLANE)

To mark two dots on the $X = 0$ face of the dice, **cube1**, we place the center of the upper one of the two disks at (0,2.875,2) along with a radius point of (0,2.875,1.375). The center of the lower disk is placed at (0,1.125,2). Both are so positioned by use of the commands **Geometry**, **Copy**, **Single**, and **Object**.

FACE WITH THREE DOTS ($Y = 0$ PLANE)

Three dots modeled by three cylindrical disks are to be placed on the $Y = 0$ face of the dice, **cube1**. The middle disk has a center point at (2,0,2) and a radius point at (2.5,0,2). This disk is then copied twice by using the **Single** option and its center as the base point. The target points should be (3,0,1) and (1,0,3).

FACE WITH FOUR DOTS ($Y = 4$ PLANE)

To place four dots on the $Y = 4$ face of the dice, **cube1**, the lower, right disk is to have a center point at (2.875,4, 2.875) and a radius point at (3.5,4,2.875). Use of the commands **Geometry**, **Copy**, **Matrix** (two rows and two columns), and **Object** enables the other three disks to be placed at the center points (1.125,4, 1.125), (1.125,4, 2.875), and (2.875,4,1.125).

FINISHED COMPOSITE SOLID — DICE

Subsequent subtractions of the required number of disks on each face from the cube complete the marking (the term *carving* may be more appropriate) of the dots necessary to develop the cube into a dice. Figure 7.4.8 is a four-way, split-screen display of the dice. Figure 7.4.9 shows a number of isometric views of the dice with the hidden lines removed. The methods used may not be the most efficient ways of constructing a dice but serve as good illustrative examples of applying the various capabilities of **SilverScreen**. Other methods such as application of **C-space** and **Paint** to be introduced later could also be employed for modeling of a dice.

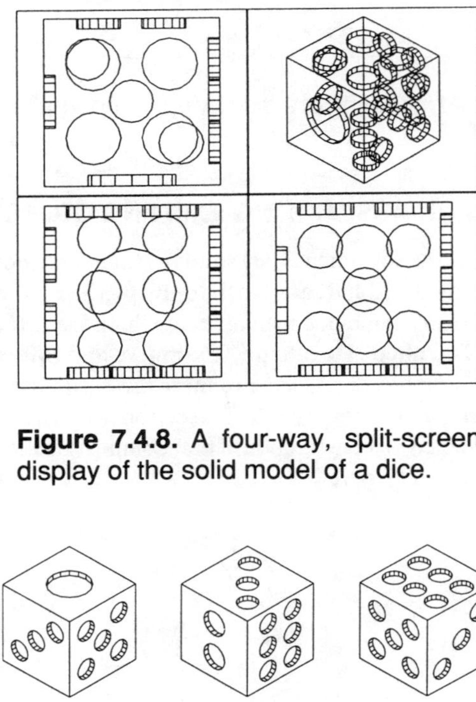

Figure 7.4.8. A four-way, split-screen display of the solid model of a dice.

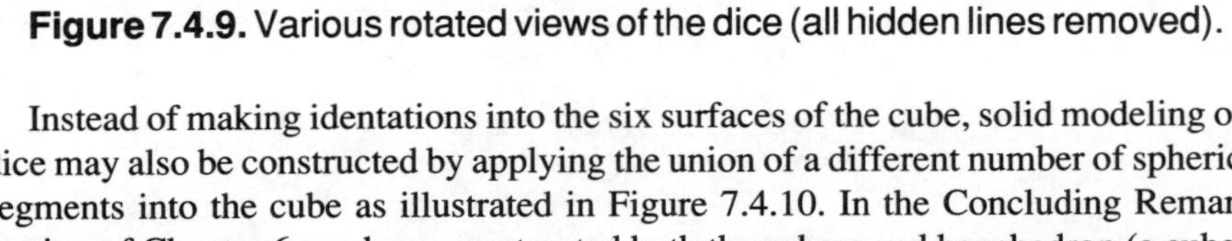

Figure 7.4.9. Various rotated views of the dice (all hidden lines removed).

Instead of making identations into the six surfaces of the cube, solid modeling of a dice may also be constructed by applying the union of a different number of spherical segments into the cube as illustrated in Figure 7.4.10. In the Concluding Remarks section of Chapter 6, we have constructed both the sphere and hexahedron (a cube is a special case of it), which are stored in the drawing file **PaoSolid.DRW**. We may therefore simply retrieve these two basic solids and modify them to meet our need here for modeling of a dice. This approach will be fully demonstrated in the next section when we discuss methods of assembling a number of basic solids that can be extracted from some existing drawing files.

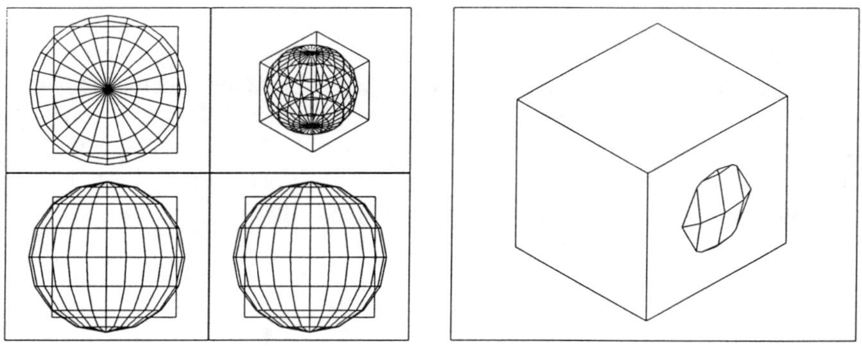

Figure 7.4.10. Solid modeling of a dice by unioning a spherical cap to mark its one-dot face.

7.5 MODIFYING AND ASSEMBLING EXISTING OBJECTS

In Section 6.8, four basic three-dimensional solids (tetrahedron, hexahedron, cylinder, and sphere) have been modeled and saved in the drawing file **PaoSolid.DRW**. A script file **PaoSolid.SF** has also been made available, in which these four solids have been given object names **Tetrahed**, **Hexahedn**, **Cylinder**, and **Sphere**, respectively. A display of Figure 6.8.2 can be generated by running this script file with the commands **Language** and **Execute** of **SilverScreen**. In this section, we will demonstrate how such basic solids can be extracted for construction of new composite solids or for assemblying into a desired combination (Figure 7.5.1).

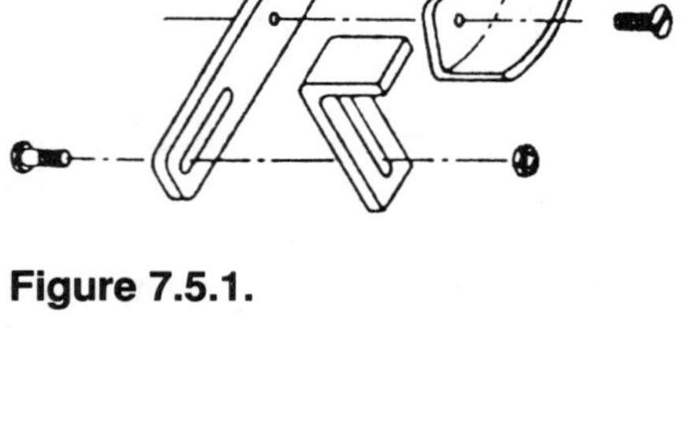

Figure 7.5.1.

An alternative way of modeling a dice previously depicted in the preceding section is by extracting the basic solids hexahedron and sphere available in the file **PaoSolid.DRW**. **SilverScreen** has an **Xcopy** command in the **I/o** menu (Figure 5.2.1) that can be applied to extract a desired object residing in a drawing file. For example, to extract the object tetrahedron saved under the name **Hexahedn** from the drawing file **PaoSolid.DRW** we enter the commands

I/o Xcopy External
Enter drawing name: **PaoSolid**
Object
Enter object name: **Hexahedn**
W-space
Enter scaling factor(s):
(Allow the scales of the object to be changed; press the **Enter** key to ask for no change.)
Enter angle of rotation(s):
(Allow the object to be rotated; press the **Enter** key to ask for no rotation.)
Mark or enter insertion point:
(Adjust the screen cursor to indicate the desired location.)

The object **Sphere** can be similarly extracted from the drawing file **PaoSolid.DRW**. The resulting drawing of the two extracted objects **Hexahedn** and **Sphere** may appear as shown in Figure 7.5.2.

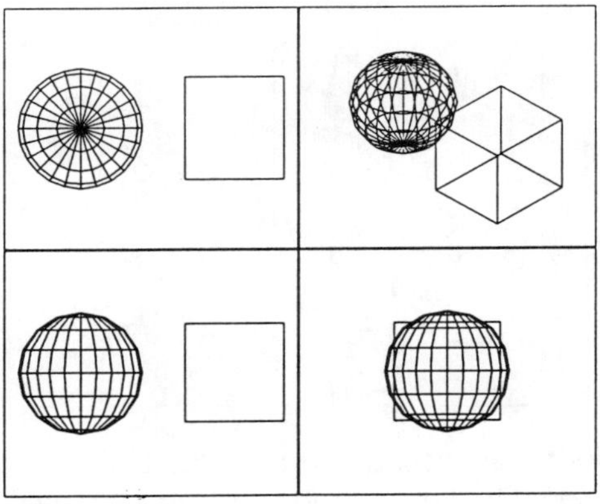

Figure 7.5.2. Two solids models: **Hexahedn** and **Sphere**.

A variety of solids can now be constructed by copying and modifying these two extracted objects, **Hexahedn** and **Sphere**. As a demonstration, let us use them to model a dice having one of its six faces marked with a spherical segment to represent the one-dotted face. If the right face of the dice is to be marked with one dot, the cube can be moved to align with the sphere as shown in Figure 7.5.3. Next, the portion of the sphere that is to the left of this right face of the cube is to be clipped away by application of the **sOlid** and **Clip** commands in the **Geometry** menu of **SilverScreen**. The right face of the cube is to be selected as the clip (*Yz*) plane and any point to the left of this plane can be used to specify the direction of clip. The resulting solid model of a dice with only one face marked is shown in Figure 7.5.4.

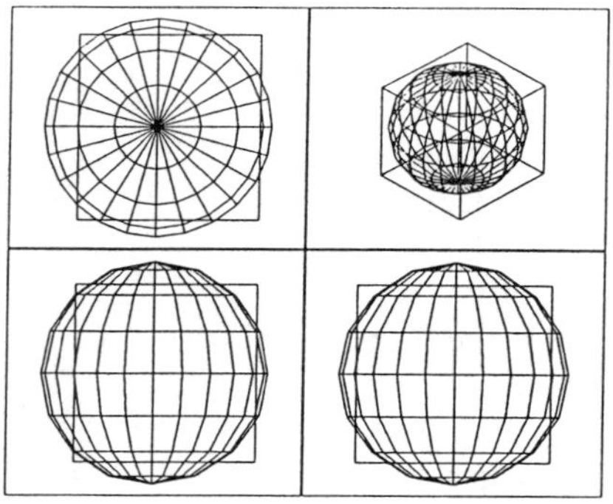

Figure 7.5.3. Align the two solids, hexahedron and sphere, before applying the clipping operation.

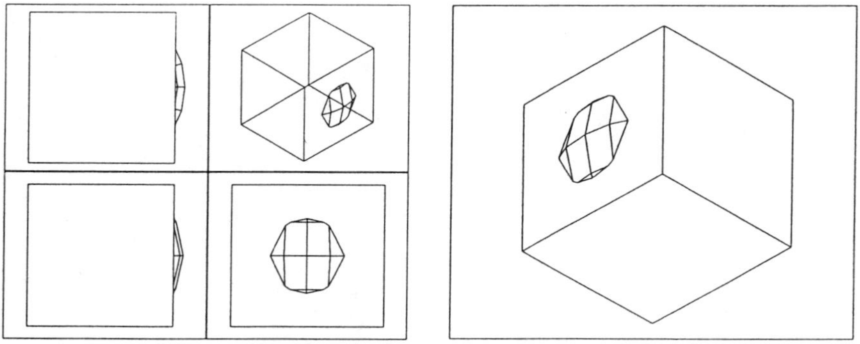

Figure 7.5.4. A four-way, split-screen display of the dice having only the on-dotted face modeled (left panel), and an isometric view having all hidden lines removed (right panel).

MODELING OF AN EXHIBIT PEDESTAL

As another example of extracting objects from an existing drawing file, we may use the objects hexahedron, cylinder, and sphere in the file **PaoSolid.DRW** to construct a solid model representing an **exhibit pedestal**. Again, using the **Xcopy** command will enable the three objects to be extracted as shown in Figure 7.5.5. Next, a copy of the cube is made and placed to the left of the original cube as shown in Figure 7.5.6. The new cube is scaled into a flat plate by entering the commands

Modify **Scale** **Object** (Select the cube on the left.)
E-Space **Relative**
Enter scaling factor(s): x2y.25z2
(The X and Z dimensions are doubled while the Y dimensions are multiplied by .25.)

The resulting collection of the four objects is shown in Figure 7.5.7. The cube on the right is also scaled with factors $x1.5$, $y.25$, and $z1.5$. It results in a new collection of four objects as shown in Figure 7.5.8. Next, the plate on the right is moved to the top of the plate on the left as shown in Figure 7.5.9. This is accomplished by using the **Modify** and **Move** commands of **SilverScreen**.

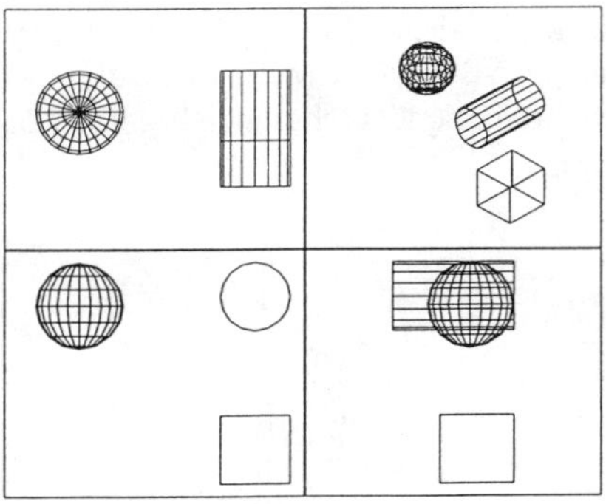

Figure 7.5.5.

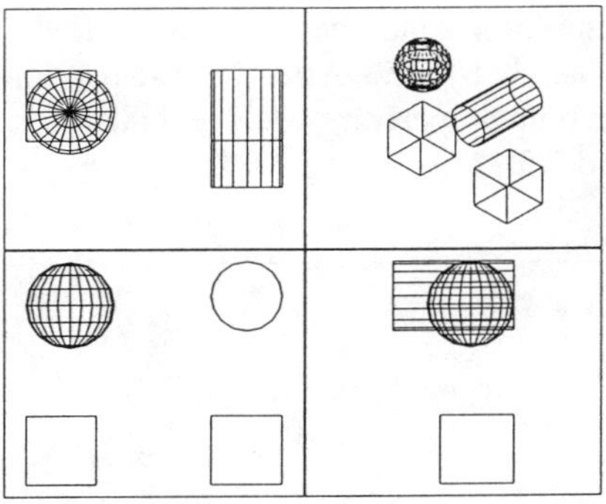

Figure 7.5.6.

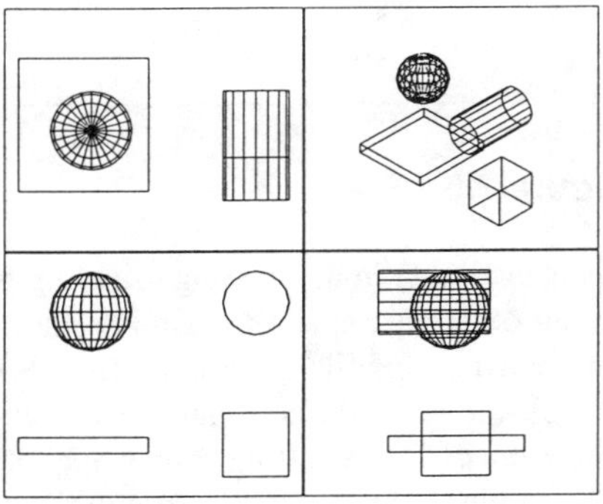

Figure 7.5.7.

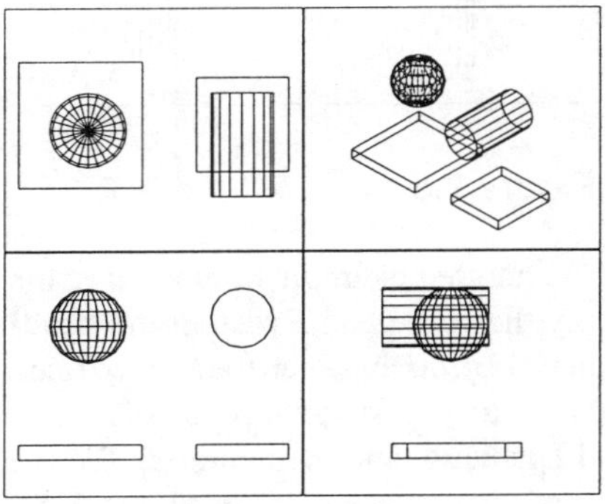

Figure 7.5.8.

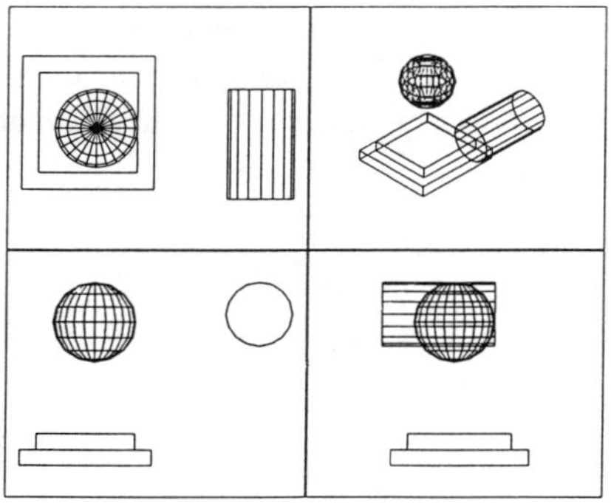

Figure 7.5.9.

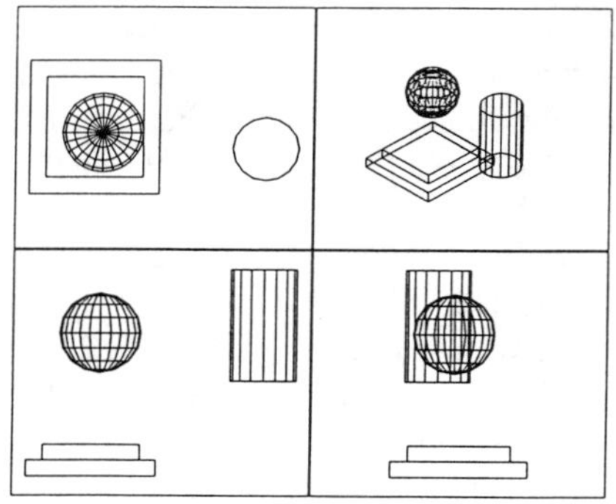

Figure 7.5.10.

Having modeled the base of the pedestal, the object **cylinder** is rotated about the *X*-axis by 90° using the commands **Modify**, **Rotate**, and **E-space**, to adapt it to become the stand of the pedestal. The resulting arrangement of the three separated objects is shown in Figure 7.5.10. To achieve a proper proportion for the different parts of the exhibition stand, the size of the cylinder is scaled by the factors $x1y1z1.5$ to increase its height by 50%. After the scaled cylinder and also the sphere (as an exhibition item) are both properly moved, the final design of the solid model is shown in Figures 7.5.11 and 7.5.12.

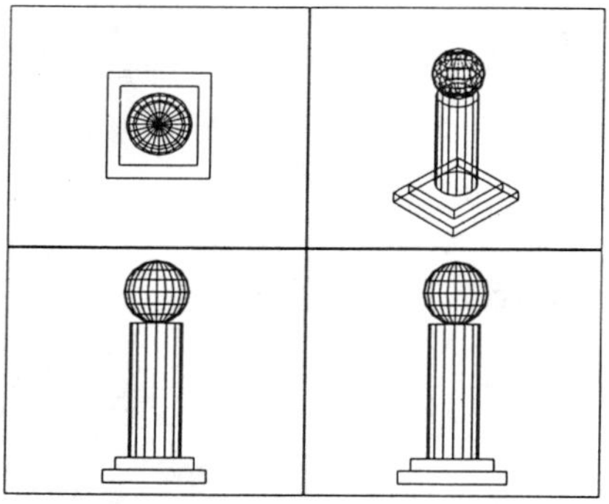

Figure 7.5.11.

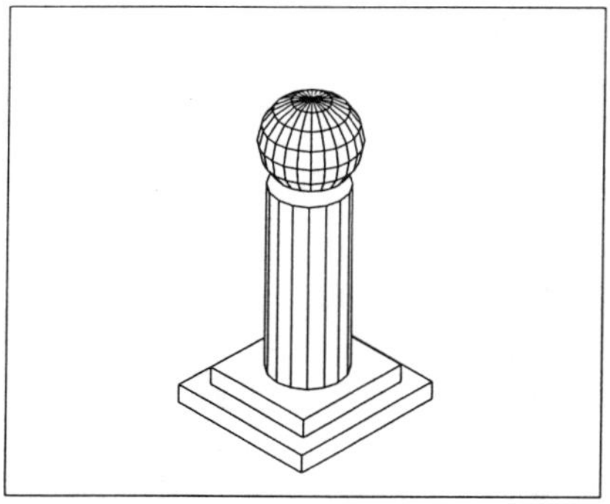

Figure 7.5.12.

As another example of extracting the objects available in an existing drawing, we apply the command **Erase** in the **Modify** menu of **SilverScreen** to remove the unwanted objects and keep the needed ones. **Booth** is a drawing file provided in the **DRW** subdirectory of **SilverScreen**; this file can be retrieved using the commands **I/o**, **Load**, and **Drawing**, and then entering the file name **Booth**. The resulting four-way, split-screen display is shown in Figure 7.5.13. The hidden-line-removed presentation of these four views can be individually achieved by exchanging the windows with the commands **Window** and **eXchange**. To remove the hidden lines, the commands

Presentation, **Hide**, **Drawing**, and **Hidden-line** need to be entered. The resulting set of four views is presented in Figure 7.5.14.

After this **Booth** has been loaded, we may extract any objects available in this assembly. For example, we have discussed the *spiral* copying method and presented a spiral stairway in Section 6.6. Here, **Booth** also has a spiral stairway. To extract this

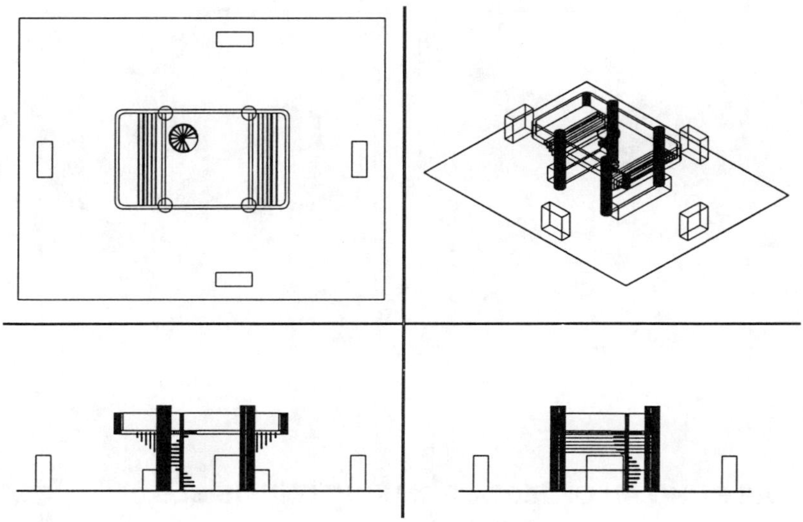

Figure 7.5.13. SilverScreen drawing file **Booth.DRW**.

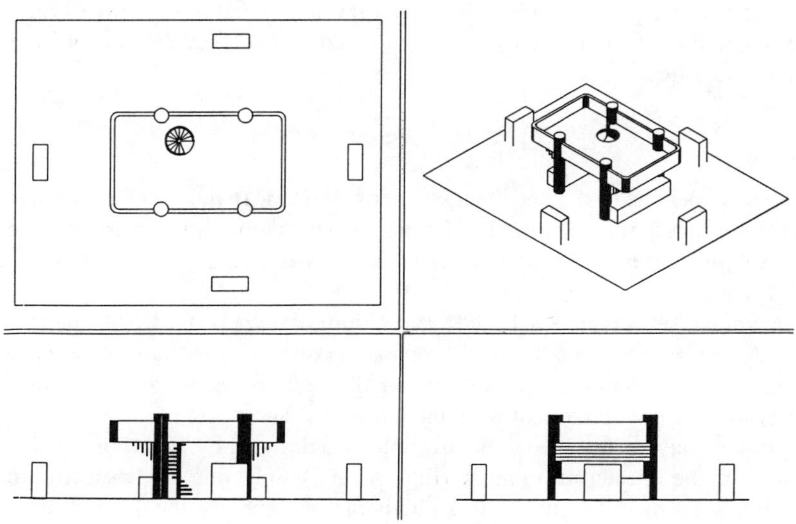

Figure 7.5.14. Hidden-lines-removed presentation.

spiral stairway, we may apply the commands **Modify**, **Erase**, **Fence**, and **Outside-plus** of **SilverScreen** to fence in only the stairway and delete all other objects that are totally or partially outside the fenced area. By so doing, an isolated object of the spiral stairway can be obtained as shown in Figure 7.5.15. A similar approach can be applied to obtain the display of the roof with four supporting columns as shown in Figure 7.5.16.

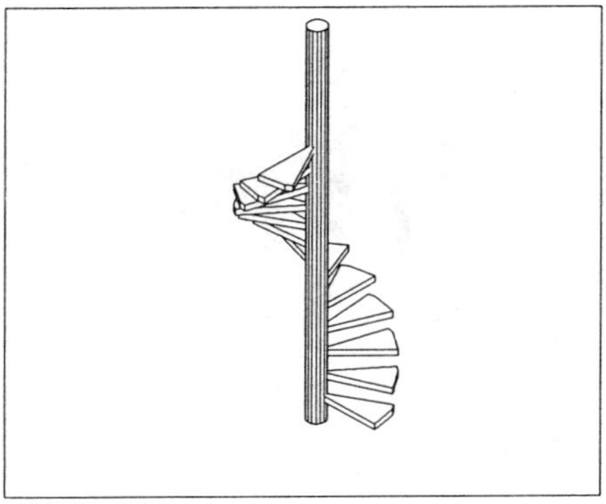

Figure 7.5.15.

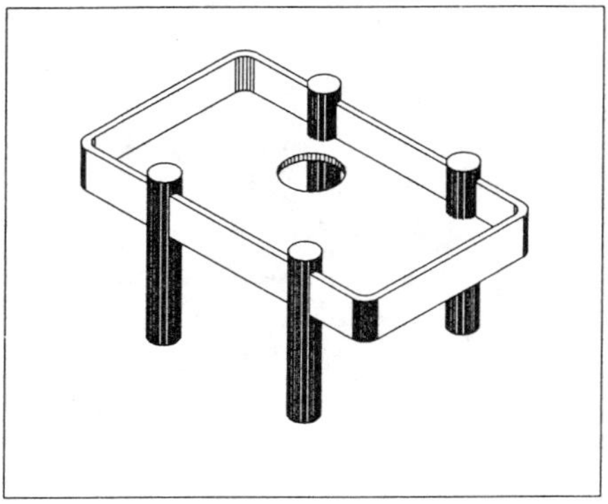

Figure 7.5.16.

7.6 DELETING AN OBJECT OR MAKING IT INVISIBLE

In Figure 7.4.7, there is a circular disk near the one-dot face of the dice. This disk, which has an object name **disk1**, was used in a copying process to generate five identical circular disks that were needed for marking the five-dot face of the dice. If we want to remove this disk from the drawing, the **Structure** menu (Figure 7.6.1) of **SilverScreen** can be applied to restructure the contents of this drawing. To **delete** the object **disk1**, we enter the following commands:

Structure Delete Object

The object **disk1** can be deleted by either using the **Pop-up** menu (**F2** option) or using the arrow keys or the pointing device (**F1** option of pick mode). To reverse the decision, the **Undelete** command in the **Structure** menu can also be applied to recover the last deleted object.

A last-drawn object can be delected by the **Undo** command in the **Utility** menu (Figure 7.6.2). All commands involved in the creation of that object are to be erased by continuously pressing the **Enter** key. When all execution steps have been erased, the **F10** key is to be pressed to exit from the **Undo** operation.

An object may be arranged to be invisible on a drawing by applying the **Visibility** command in the **Presentation** menu (Figure 2.5.3). Selecting the **Disable** command and pointing to the object that is to be invisible will achieve the desired result. The

STRUCTURE

Attribute
Create
Delete
Group
gOto
Join
List
Newname
Purge
Relocate
Tree
Undelete

Figure 7.6.1.

UTILITY

Clear
Execute
Help
Linestyle
Multi-window
Penstyle
Startup
Text-edit
Undo
uNits

Figure 7.6.2.

object can be made visible again by selecting the **Enable** command. We can make the sphere in Figure 7.5.12 invisible and have a hard copy of the exhibit stand as shown in Figure 7.6.3 by entering the commands **Presentation**, **Visibility**, **Disable**, and **Object** and using the pointing device or the arrow keys to select the sphere.

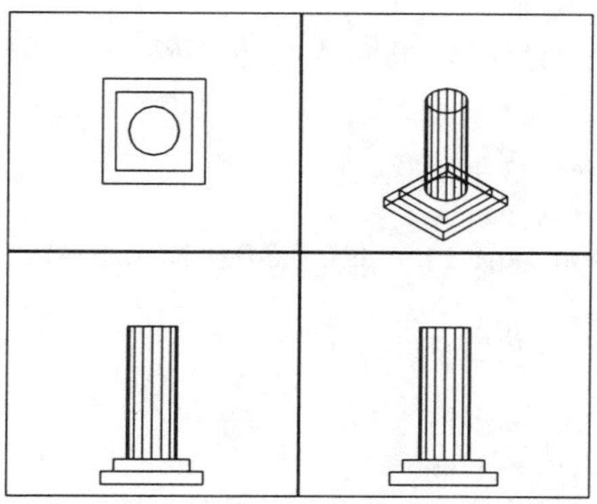

Figure 7.6.3.

7.7 CONCLUDING REMARKS

As we gradually generate a variety of three-dimensional solid models by the methods introduced in Chapter 6 and this chapter, it is a good practice to save them in drawing files for the purpose of serving the needs of forming other composite solids. For example, the right hexagonal prism and cone are often encountered in construction of solids such as the head portions of bolts and screws. We may therefore want to include them in an extended drawing file **Pao3dSdX.DRW** such that the collection will have six objects (Figure 7.7.1).

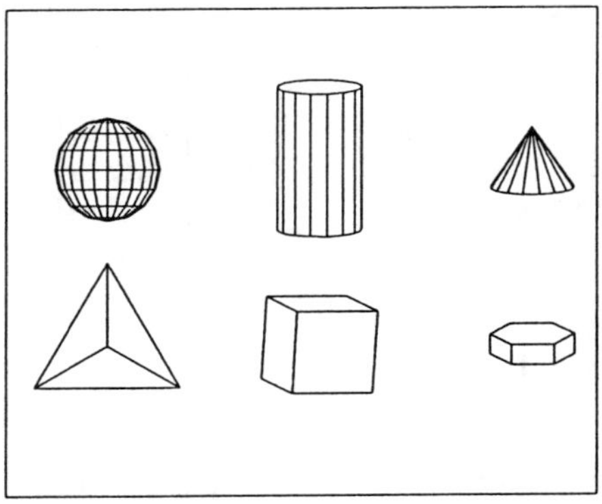

Figure 7.7.1.

The script file **Pao3dSdX.SF,** which can be run to generate the extended drawing file **Pao3dSdX.DRW** by using the **Execute** command in the **Language** menu, is listed below.

(First part contains statements in the file **PaoSolid.SF** — see Section 6.8.)

```
at 0,0,0
draw equilateral vertex 3 delta 0,4,0
window split four-way drawing
sweep linear polygon \obj1.1 points 0,0,0 0,0,3.5 scale 0
 name tetrahed
window split four-way drawing
      at 10,0,0
draw equilateral vertex 4 delta 2,2,0
window split four-way drawing
sweep linear polygon \obj2.1 points 12,–2,0 12,–2,4.25
 scale 1 name hexahedn
at 10,9,0
draw circle radius 2,0,0
display refine circle sweep 16
sweep linear cirlipse \obj3.1 points 12,9,0 12,9,7 scale 1
 name cylinder
window split four-way drawing
at 0,8.5,2.5
draw circle radius 2.5,0,0
display refine circle sweep 16
```

```
sweep circular cirlipse \obj4.1 axis 0,9,2.5 0,8,2.5
 angle 180 faces 12 horizontal 15 vertical 15 name sphere
(Draw a right hexagonal prism and name it "prismhex.")
at 20,0,0
draw equilateral vertex 6 delta 0,2,0
window split four-way drawing
sweep linear polygon \obj5.1 points 21.75,0,0 21.75,0,1
 scale 1 name prismhex
(Draw a cone and name it "cone.")
at 20,9,0
draw circle radius –2,0,0
display refine circle sweep 16
sweep linear cirlipse \obj6.1 points 20,9,0 20,9,3 scale 0
 name cone
```

This script file is obtained by applying the commands **Remember** and **Append** in the **Language** menu of **SilverScreen** to the file **PaoSolid.SF**. **Polygon**, **Equilateral**, **Vertex**, **Circle**, and **Radius** commands in the **Draw** menu have been applied to generate the objects **prismhex** and **cone**, respectively. A major difference in the sweeping process of the two solids, **prismhex** and **cone,** is the scaling factors — the values 1 and 0 are used for them, respectively.

The examples presented in this chapter deal with the solids that can be easily constructed using the world space (**W-space**). A cube with faces parallel to the coordinate axes *X*, *Y*, and *Z*, and a circular cylinder having its axis parallel to any one of the coordinate axis are constructed by simple sweeping processes when sweeping axes are easy to define. If an oblique plane is defined and some other objects are to be stacked onto that plane, it is then of advantage to define a new set of coordinate system having two of its axes placed on this oblique plane. In such an arrangement, the other objects can then be defined much easier than in the W-space. This new coordinate system, which defines a **Construction space** (**C-space**), is the subject of Chapter 8.

Various ways of presenting a three-dimensional composite solid by application of the **Presentation** menu of **SilverScreen** are the topics in Chapter 9. Shading and other techniques will be introduced that enable us to visually clarify whether the cylinder shown in Figure 7.2.1 is a hole or an insert.

In Chapter 10, we will discuss the combined use of **SilverScreen** commands and the statements of **BASIC** and **C** languages for generation of various solids. Such an arrangement enables three-dimensional solids to be defined and constructed using mathematical equations in a **BASIC** or **C** program.

PROBLEMS

7.1 Construct the bumper support and post cap shown in Figure P7.1 by Boolean union and subtraction.

7.2 Construct the anchor bracket shown in Figure P7.2 by Boolean union and subtraction.

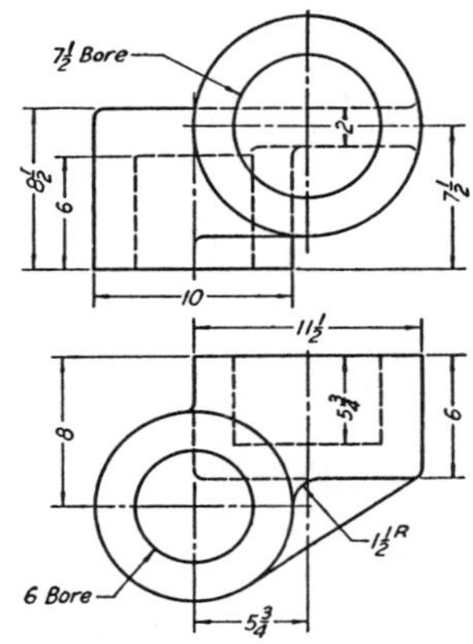

Figure P7.1.

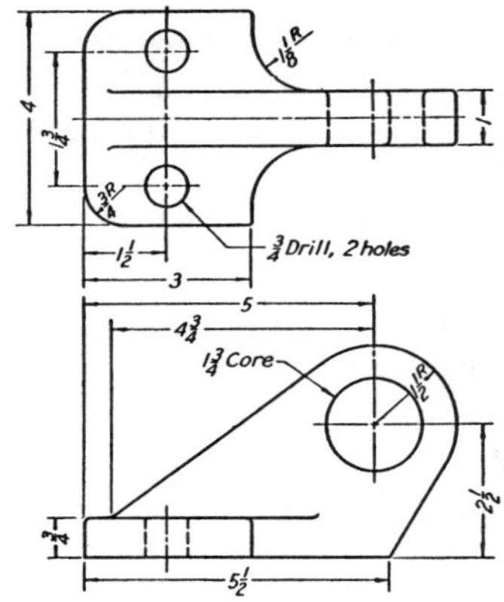

Figure P7.2.

7.3 How should **SilverScreen** be applied to obtain a hard copy of Figure P7.3 by rotating the dice and exchanging the windows shown in Figure 7.4.8?

7.4 How should **SilverScreen** be applied to obtain a hard copy of Figure P7.4 by rotating the dice and exchanging the windows shown in Figure 7.4.8?

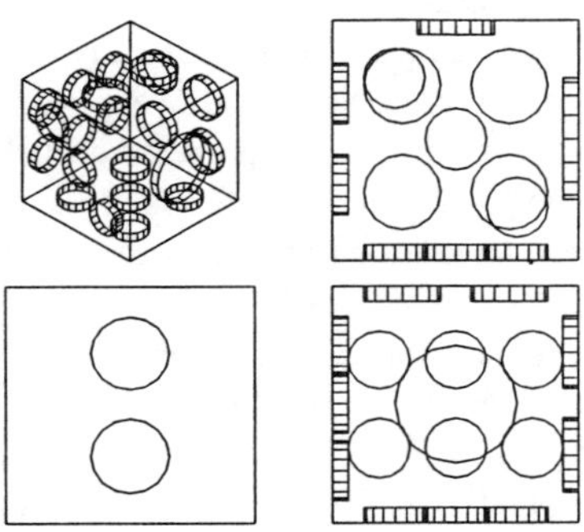

Figure P7.3.

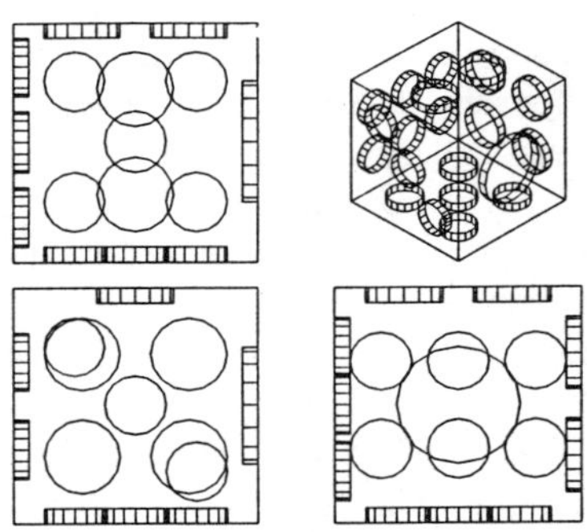

Figure P7.4.

7.5 How should **SilverScreen** be applied to obtain a hard copy of Figure P7.5 by rotating the dice and exchanging the windows shown in Figure 7.4.8?

7.6 How should **SilverScreen** be applied to obtain a hard copy of Figure P7.6 by requesting only a bordered printout of the current window shown in Figure P7.5?

7.7 A solid model for a gear with 18 teeth is shown in Figure P7.7. First draw a wedge and then linearly sweep it into a three-dimensional tooth. Next a circular disk is to be constructed as the base for mounting the tooth at 18 different locations by a repetitive Boolean-union process.

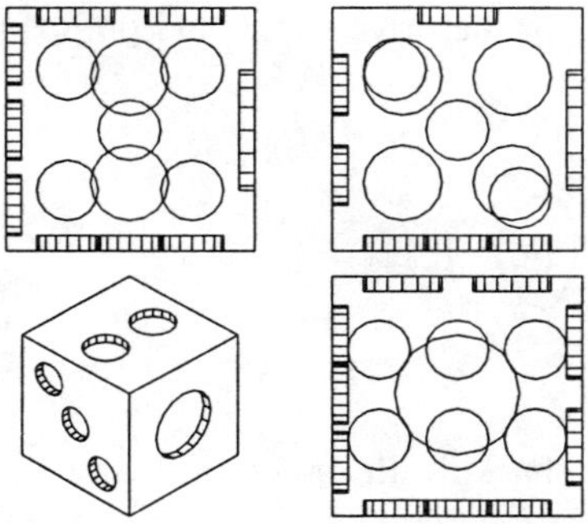

Figure P7.5.

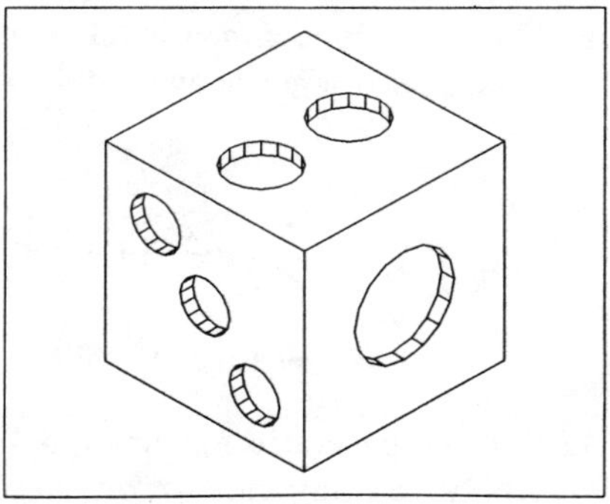

Figure P7.6.

7.8 Use Boolean subtraction to construct a Woodruff keyseat as shown in Figure P7.8 by removing a half circular disk from a circular cylinder.

7.9 How should **SilverScreen** be applied to obtain a hard copy of Figure 7.4.1?

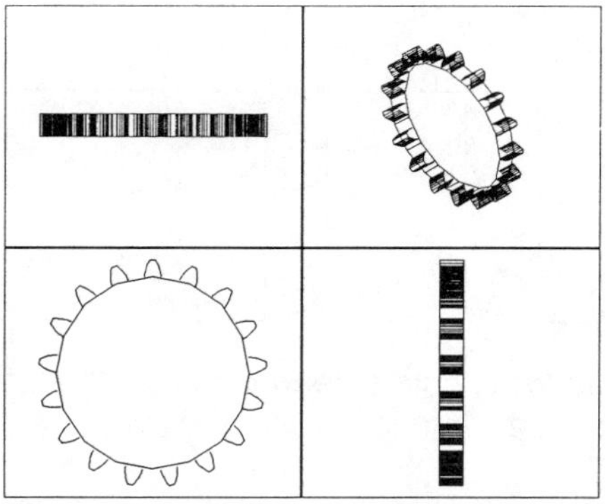

Figure P7.7.

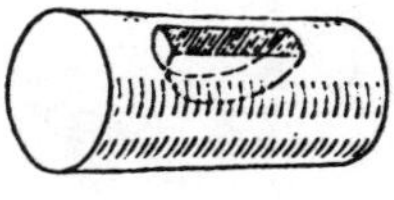

Figure P7.8.

7.10 How should **SilverScreen** be applied to obtain a hard copy of those rotated, hidden-line-removed views of the dice shown in Figure 7.4.9?

7.11 Construct three-dimensional solid models for the stove bolts shown in Figure P7.11.

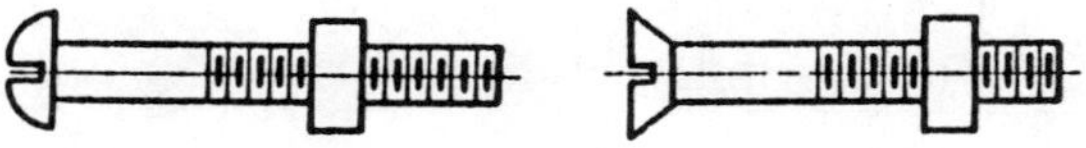

Figure P7.11.

7.12 Construct three-dimensional solid models for the drive screw and philips recessed head screw shown in Figure P7.12.

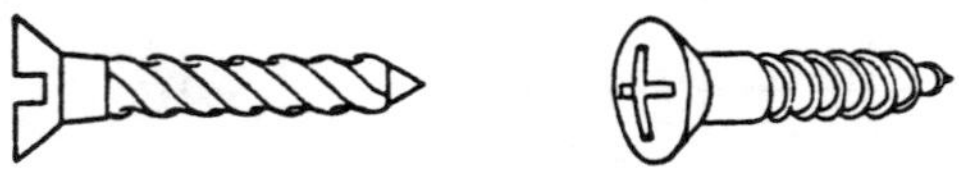

Figure P7.12.

7.13 The star generated in Section 6.3 has a flat back as illustrated in Figure 6.3.3. Apply the **Copy** and **mIrror** commands in the **Geometry** menu to generate a full star as shown in Figure P7.13.

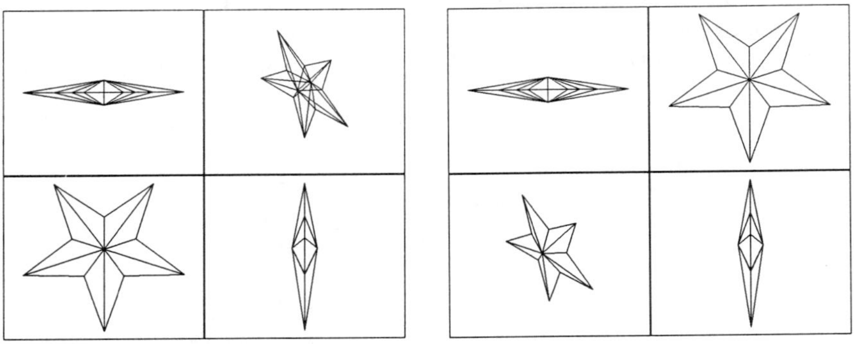

Figure P7.13. Four-way, split-screen displays of a star (the right panel shows the hidden-lines-removed version of the isometric view).

7.14 Three different plates have been constructed in Problems 6.4, 6.6, and 6.8 in Chapter 6. Apply the commands **Move** and **Rotate** or **A-rotate** in the **Modify** menu to assemble them into a composite structure as shown in Figure P7.14.

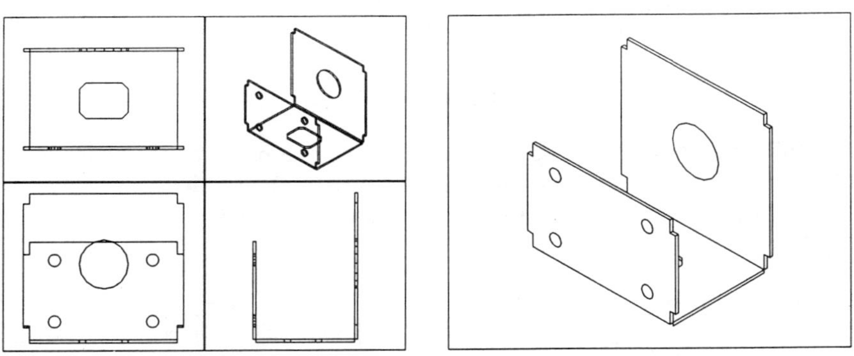

Figure P7.14.

7.15 Construct a solid model for the housing plate shown in Figure P7.15.

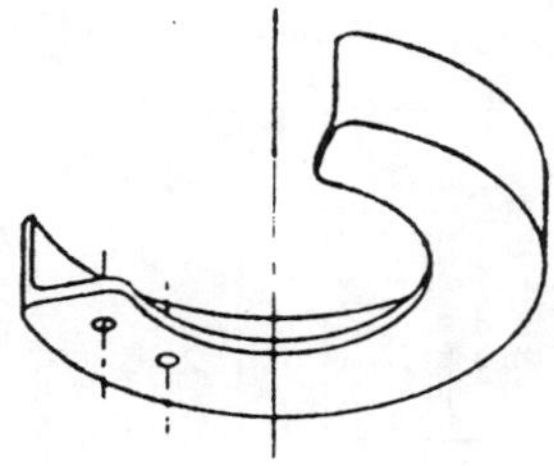

Figure P7.15.

7.16 Constructed the mounting block shown in Figure P7.16 by utilizing the basic solids available in the drawing file **PaoSolid.DRW**.

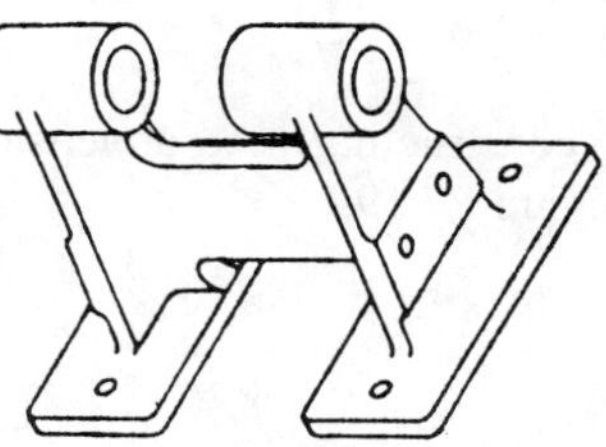

Figure P7.16.

7.17 Construct a three-dimensional solid model for the bolt having a hexagonal head shown in Figure P7.17.

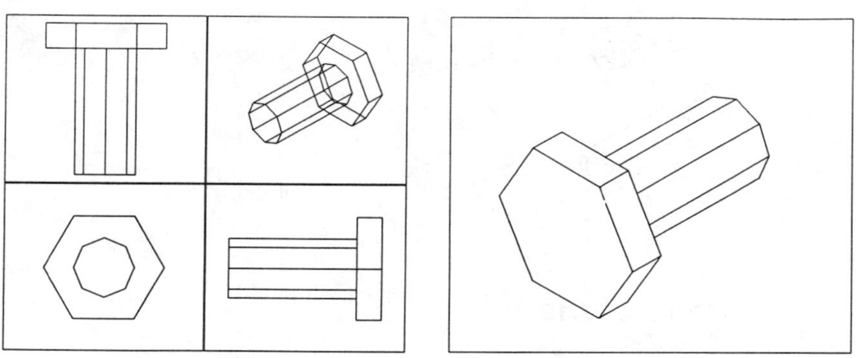

Figure P7.17.

7.18 Apply the **Copy** and **Sweep** commands of **SilverScreen** effectively for modeling a three-dimensional "window" structure shown in Figure P7.18.

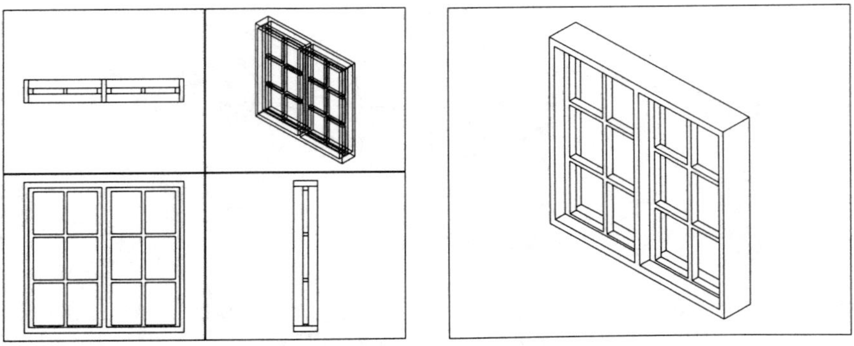

Figure P7.18.

7.19 Apply **SilverScreen** for constructing three-dimensional solid models of those RF connectors shown in Figure P7.19.

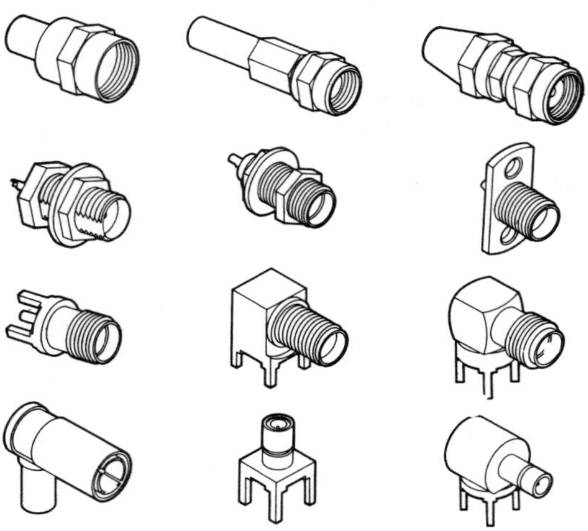

Figure P7.19.

7.20 Construct three-dimensional solid models for the objects shown in Figure P7.20.

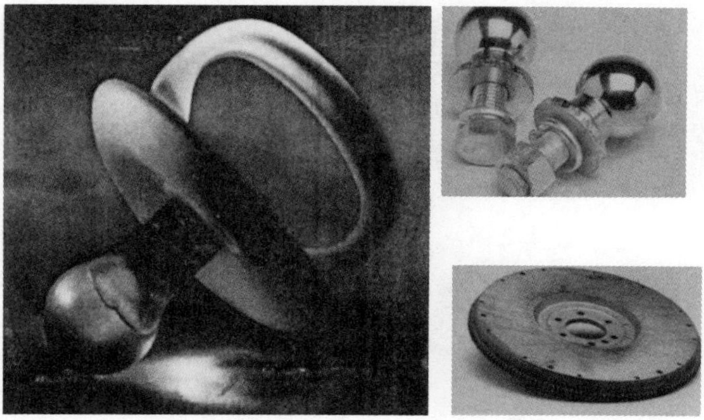

Figure P7.20.

7.21 Instead of the arrangement as shown in Figure 7.5.4, rotate the sphere before it is combined into the cube so that either its north or south pole will be shown as the "dot" on the dice.

8 Construction Space and Auxiliary Views

8.1 INTRODUCTION

SilverScreen provides the convenience of having three sets of coordinate systems for construction of three-dimensional solid models. They are referred to as the **W-space**, **C-space**, and **E-space** for the world, construction, and entity spaces, respectively. World space is a fixed coordinate system having the *X*-axis and *Y*-axis along the horizontal and vertical directions of the display screen and the *Z*-axis perpendicular to the screen. World space remains stationary and unchanged.

Construction space is a coordinate system having *X*- and *Y*-axes located on a selected plane to facilitate the creation of primitives needed in construction of three-dimensional solids. For example, if one needs to cut a cylindrical hole perpendicular to the inclined plane ABCD shown in Figure 8.1.1 it is more convenient to specify the location of the axis of the hole by having the *X*- and *Y*-axes on the inclined plane. The

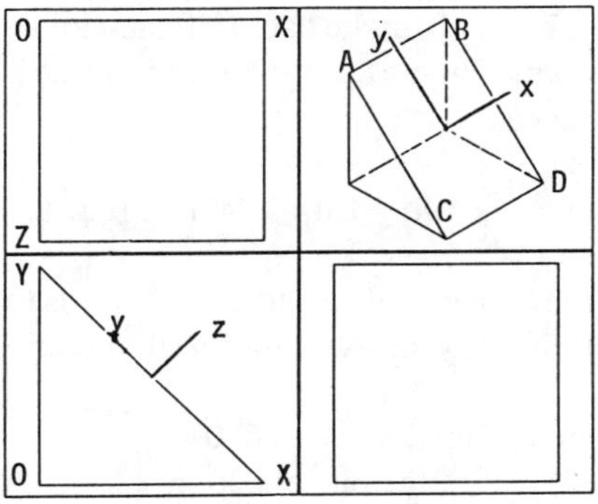

Figure 8.1.1.

Z-axis is uniquely defined by the right-hand triad consideration. It will be much more cumbersome to define the shape and size of the cylindrical hole by referring to the world space (using the *X*-, Y-, and Z- axes in Figure 8.1.1).

In this chapter, we will discuss how a construction space can be defined in various ways and in terms of the W-space coordinates by applying the **Construct** menu (Figure 8.1.2) of **SilverScreen**. **Alt** and **W** keys can be pressed to toggle the display of the coordinates from W-space to C-space or vice versa in the **Status Area**. As shown in Figure 8.1.3, as the background color of the area showing the changes in the values of the three coordinates, the exchange of the two spaces is made.

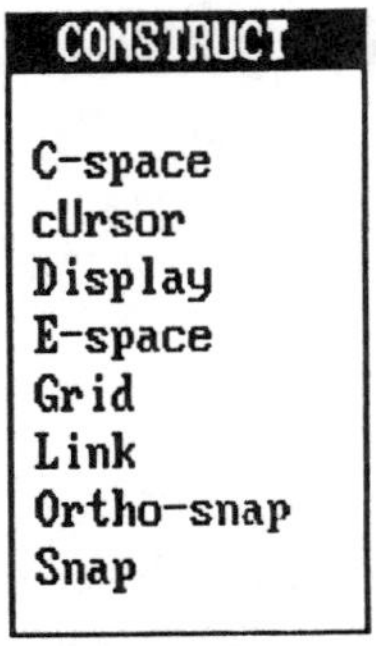

Figure 8.1.2.

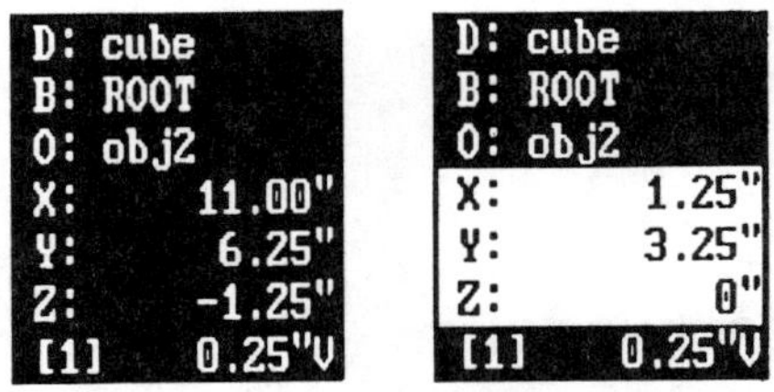

World Space Coordinate Entry

Construction Space Coordinate Entry

Figure 8.1.3.

Entity space is the space used in construction of an entity. A drawing may have many entities and each entity has its own E-space. This arrangement makes it possible to move, rotate, and scale any entity relative to the others. In the following sections, we introduce various ways of applying these three spaces with the aid of many examples.

8.2 AN EXAMPLE OF USING C-SPACE

Let us consider the problem of cutting a circular, cylindrical hole perpendicular to the plane that divides a cube of size 6′×6′×6′ into two equal halves as shown in Figure 8.2.1. The W-space has the origin (0,0,0) at the back, left, and lower corner of the cube. The *X*-, *Y*-, and *Z*-axes are shown in Figure 8.2.1a. To generate the half cube shown in Figure 8.2.1b, we use the commands

Geometry sOlid Clip Object (cube) Plane

The cube is selected as the clipped object and a plane needs to be defined for clipping. Three points are to be specified by the user to define the clipping plane, such as follows:

Mark or enter first point of the plane : 0′,6′,0′
Mark or enter second point of the plane: 0′,6′,6′
Mark or enter third point of the plane : 6′,0′,6′

They are the points A, B, and C shown in Figure 8.2.1a. Next, one must specify which half of the cube is to be clipped away by responding:

Mark or enter point to indicate direction of
clipping: 6′,6′,0′

The entered coordinates specify the point E in Figure 8.2.1a.

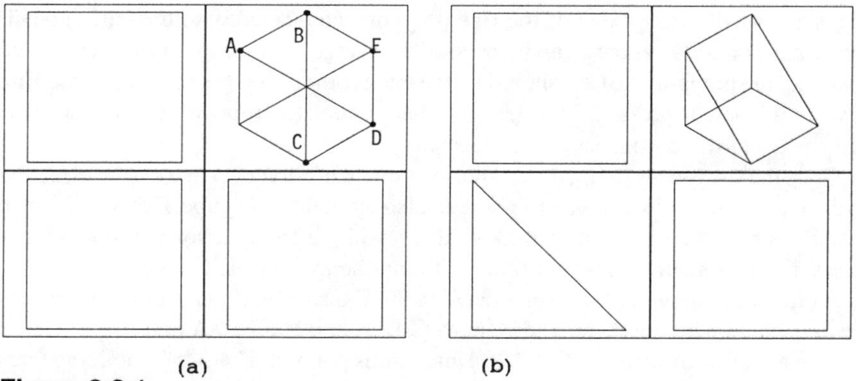

Figure 8.2.1.

The next two questions and responses are concerned with how the cutting plane should be displayed:

Enter color of new surface: 8 (black)
Enter pattern for new face: (none)

Figure 8.2.1b is the resulting four-way, split-screen display of the remaining half-cube.

Next, we cut a circular, cylindrical hole through the half cube. This hole is to be cut perpendicular to the inclined plane ABC shown in the upper right panel of Figure 8.2.2. For this particular demand, it is advisable to request a C-space construction. That is, we should ask to place the origin of a new coordinate system at the center of the plane ABC and *X*- and *Y*-axes to be directed parallel to the sides of the polygon; the circular, cylindrical hole will then have its axis along the *Z*-axis. This circular, cylindrical hole is then much easier to define than the case when the W-space is used. The commands to make such a request for a C-space construction are

Construct C-space Align Primitives

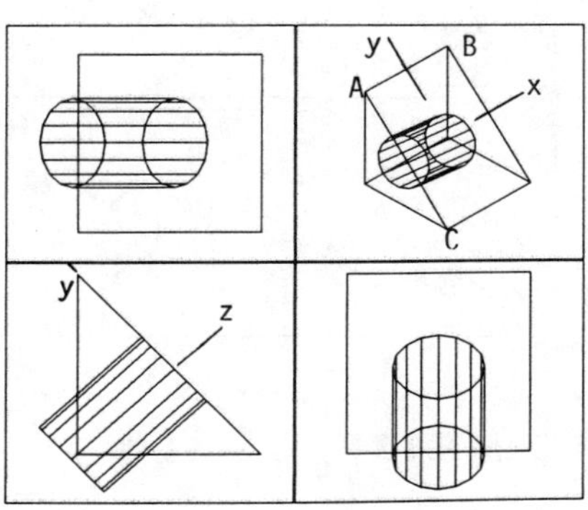

Figure 8.2.2.

The screen will prompt one of the five polygons connected with the half-cube. The arrow keys on the keyboard can be pressed to change the prompt to another polygon. To select the polygon ABC, it should be the one prompted on the screen and the **Enter** key should then be pressed. This commits the decision to the polygon, on which the *X*- and *Y*-axes of the C-space are to be located.

Once the C-space is defined, the **Window** menu of **SilverScreen** can be effectively utilized to request a four-way, split-screen display of the half-cube. Cross-hair cursors will appear in the three orthogonal windows while a triad cursor will appear in the upper, right, isometric-view window, showing the axes of the C-space.

A circular cylinder is next constructed in the C-space by first generating a circle on the *X–Y* plane with the commands **Draw**, **Circle**, and **Radius**. A pointing device can be used to define the center (3′, 4′3″,0′) and radius point (4′6″,4′3″,0′), noticing that all values of the coordinates refer to the C-space. As the pointing device gradually moves away from the defined center, one witnesses on screen a circle that is expanding radially outward on the ABC plane in the isometric-view window. The cylinder is next formed by a sweeping process using the commands **Geometry**, **Sweep**, **Linear**, and **Object** (selecting the newly constructed circle), base point (1′6″,4′3″,0′), target point (1′6″,4′3″,-4′6″), and a scaling factor equal to 1. The combined drawing of the half-cube (object name **cube1**) and the newly generated cylinder (object name **Cylind**) is shown in Figure 8.2.2.

To cut a circular, cylindrical hole having the size and shape of the object **Cylind** into the half-cube, we apply the commands **Geometry**, **sOlid**, and **Difference**, and specify the object **cube1** as the primary object and the object **Cylind** as the selected object to be subtracted. The resulting solid is shown in Figure 8.2.3. To make a better presentation with the hidden-line removed, the four panels of Figure 8.2.4 have been obtained by applying the **Hide** command to the current window and then exchanging one by one the other three panels with the current window in Figure 8.2.3.

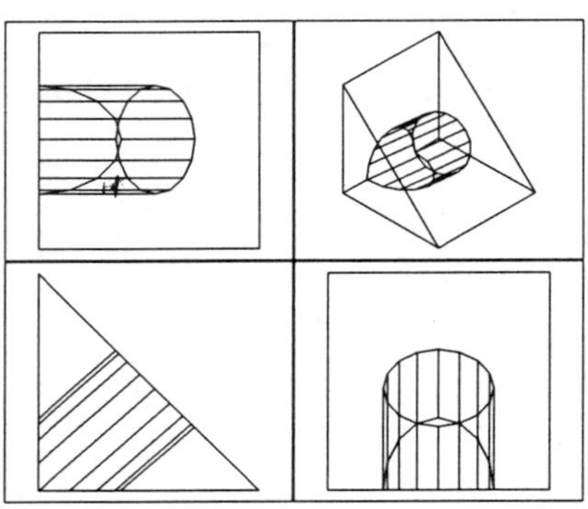

Figure 8.2.3.

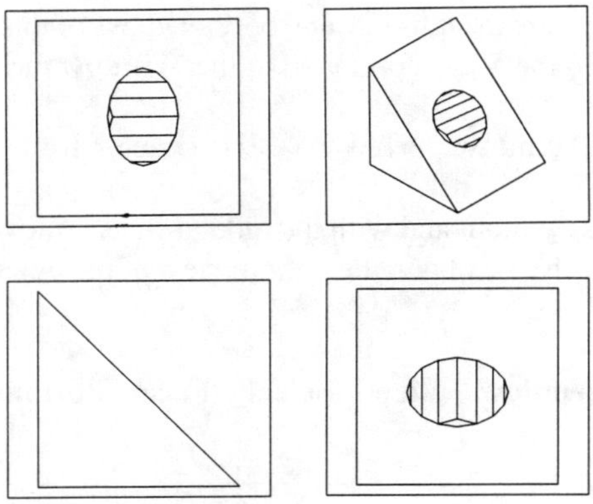

Figure 8.2.4.

8.3 AUXILIARY VIEW

To describe the true shape of a slanting plane such as the plane ABC shown in Figure 8.2.2, it is often advisable in practical drafting to take an **auxiliary view** perpendicular to this plane. Let us illustrate the procedure of using the **SilverScreen** software for taking auxiliary views, again by use of the half-cube with a circular, cylindrical hole previously shown in Figure 8.2.3. First, a **six-way**, split-screen display of this object can be obtained as shown in Figure 8.3.1 using the commands

Window Split Six-way Object

The top three panels shown in Figure 8.3.1 are the left-side, top, and right-side view of the object, respectively. The bottom row has a middle panel that is the **current window** and shows the front view whereas the other two are the isometric views of the object.

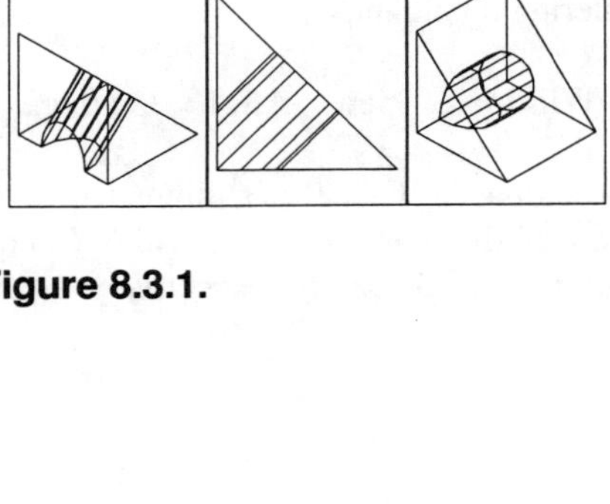

Figure 8.3.1.

Of particular interest are the left-side and bottom views of this object. To obtain the former, we can applied the **View** command in the **Window** menu as follows:

Window View Select Face Left

The resulting displays, without and with the hidden-lines removed, are shown in Fig. 8.3.2a and b, respectively. To obtain the bottom view of the object as shown in Figure 8.3.3, we enter

Window View Select Face Bottom

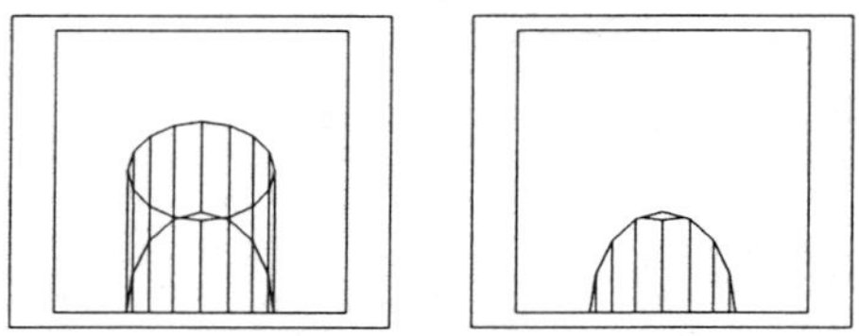

(a) (b)

Figure 8.3.2.

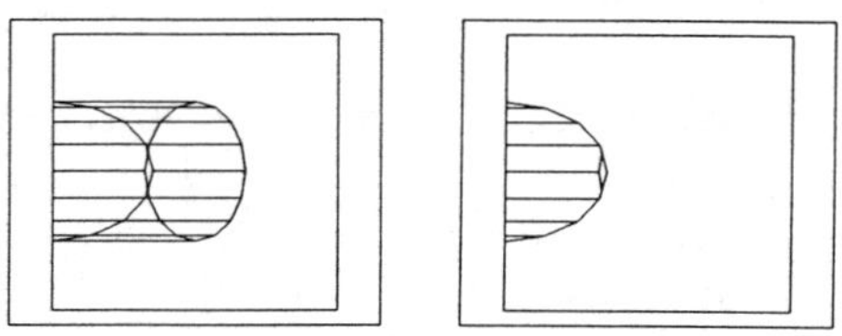

(a) (b)

Figure 8.3.3.

If we give the details of what is on the slanted plane, a right-side view of the object shown in Figure 8.3.4a will not provide the true shape even with the hidden-line removed as evidenced by Figure 8.3.4b. It is necessary to ask for an **auxiliary** view of that slanted plane by entering commands

Window View Align C-space

since we happen to know that the C-space is aligned with the slanted plane. The resulting auxiliary views of the slanted plane, without and with the hidden-line removed, are shown in Figure 8.3.5a and b, respectively. Not using the C-space, the

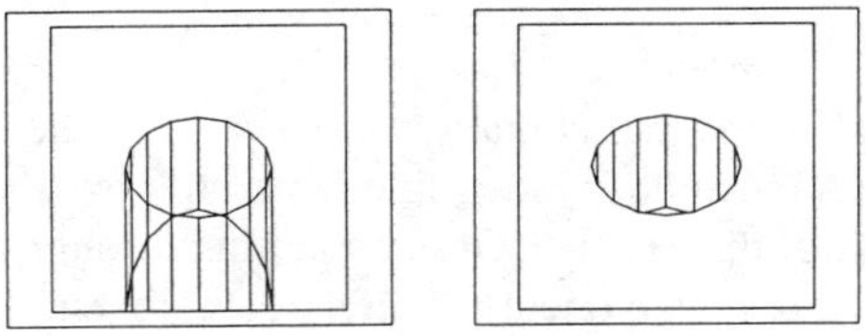

(a) (b)

Figure 8.3.4.

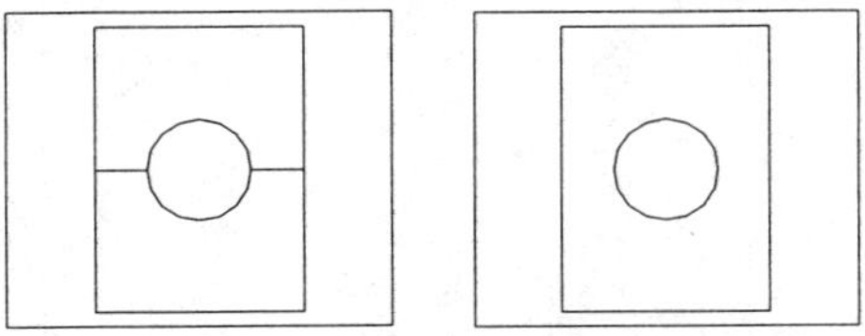

(a) (b)

Figure 8.3.5.

same result can be obtained by use commands

Window View Align Polygon (select the slanted one)

We may also effectively use the isometric views shown in the six-way, split-screen display of the object (Figure 8.3.1) to describe where the circular, cylindrical hole has been cut in the half-cube. For example, we can exchange the isometric view in the lower, left panel of Figure 8.3.1 into the current window and then use the **A-rotate** command in the **Modify** menu to rotate this view by −15°. The resulting display is shown in Figure 8.3.6a. When the hidden lines are removed, Figure 8.3.6b clearly indicates where the circular, cylindrical hole is cut through the half-cube.

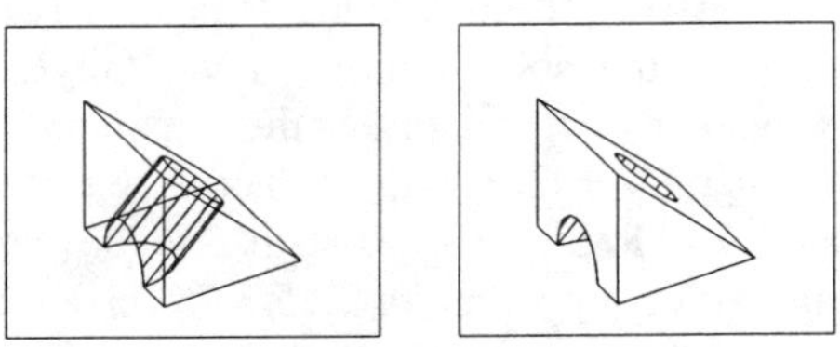

(a) (b)

Figure 8.3.6.

8.4 SECTION VIEW

For three-dimensional solids that contain cavities such as the coffee mug shown in Figure 8.4.1, even with the aid of hidden-line removal it is still difficult to tell about the sizes and locations of the cavities. Only by cutting through the solid and taking a **sectional** view of the cavity can one completely comprehend the geometric characteristics of the cavity. For the case of the coffee mug, its depth can be seen if it is cut right through the plane of symmetry.

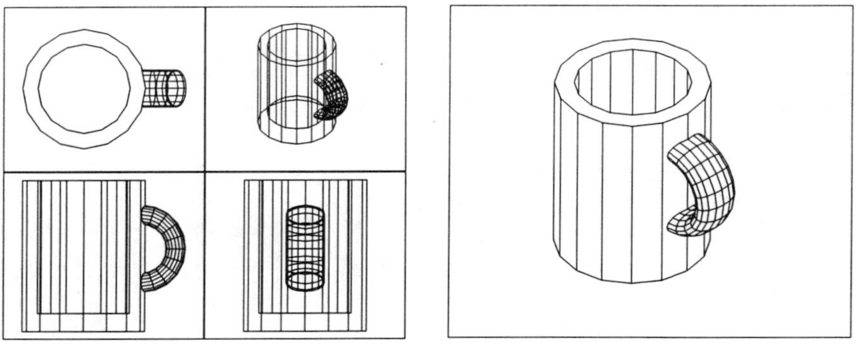

Figure 8.4.1.

To obtain a sectional view of a three-dimensional solid, the **Clip** command of **SilverScreen** can be effectively utilized. To illustrate the procedure, let us construct the coffee mug first and then cut away a half of it by application of the **Clip** command. The mug shown in Figure 8.4.1 has a height of 3-3/4 in. and an outside diameter of 3 in. and inside diameter of 2-1/4 in. The cavity has a depth of 3-1/4 in. The handle of donut shape has a radius of 7/8 in. and an elliptical cross section whose major and minor radii are equal to 1/2 in. and 3/8 in., respectively.

To construct the body of the mug, one may first draw a circle using the outside radius of the mug, which is 1-1/2 in. and the commands **Draw**, **Circle**, and **Radius**. There is by a linear sweep of this circle into a circular cylinder with a length equal to the height of the mug, 3-3/4 in. The **SilverScreen** commands involved are **Geometry**, **Sweep**, **Linear**, and **Circlipse**. Give this solid an object name **Obj1**. A second cylinder of smaller radius can be generated by first applying the **Copy** command of **SilverScreen**; that is given the object name **Obj2**. Next, we modify by applying the commands **Modify**, **Scale**, **W-Space**, and **Relative**. The two axes in the plane parallel to the cross section need to be scaled down by a ratio of 2.25/3.75 (inside diameter over outside diameter of the mug) while the scale for the third axis should remain equal to 1.

The cut-away operation of object **Obj2** from object **Obj1** is carried out by applying the commands **Geometry**, **sOlid**, and **Difference**. **Obj1** and **Obj2** should be selected as the primary and secondary objects, respectively. Furthermore, **Obj2** should be positioned correctly relative to **Obj1** so that the resulting solid is an accurate representation of the **body** of the mug with a cavity depth of 3-1/4 in. **SilverScreen** will automatically assign a new name **Obj3** to the newly derived object.

To add a **handle** to the body of the mug, we again apply the sweeping procedure of **SilverScreen**. First, an ellipse is to be drawn by using the commands **Draw** and **Ellipse.** The major and minor axes of the ellipse should have the values of 1/2 and 3/8 in., respectively, which are the cross-sectional dimensions of the handle. This ellipse is next used in a circular, sweep process to form a half donut. The commands **Geometry**, **Sweep**, **Solid**, and **Circular** are to be applied, having an axis of rotation at a distance of 7/8 in. measured from the center of the ellipse and parallel to its major axis. The angle of sweep is 180°. This half donut is next properly rotated and then placed onto the outside of the mug at a location that is 1-5/8 in. from the top of the mug measured to the mid-section of the handle. The left panel of Figure 8.4.1 shows a four-way, split-screen display of the constructed mug whereas the right panel is an enlarged, isometric view of the mug with the hidden lines removed. Even the latter display cannot reveal how deep the mug is. This calls for a sectional cut of the mug to give a description of the cavity.

The top view of the mug shown in the left panel of Figure 8.4.1 indicates that we may use a plane parallel to the *X–Y* plane to cut through the mug. If the front half of the mug is removed, then it is easy to visualize how deep the mug is. This **Clipping** process can be accomplished by use of the **SilverScreen** commands and by responding to the prompted questions as follows:

Geometry sOlid Clip Object (Obj1)
Xy-plane
Mark or enter location of z-coordinate:
(move cursor in the top view so that the horizontal hairline
passes the center of the circles.)
Enter point to indicate direction of clipping:
(move the horizontal hairline down to any location to
specify that the front half of the mug is to be clipped.)
Select or enter surface color for new faces: 15 (white)
Enter pattern: (none)

The resulting half mug is shown in Figure 8.4.2. The right panel clearly shows the details of the interior of the designed mug.

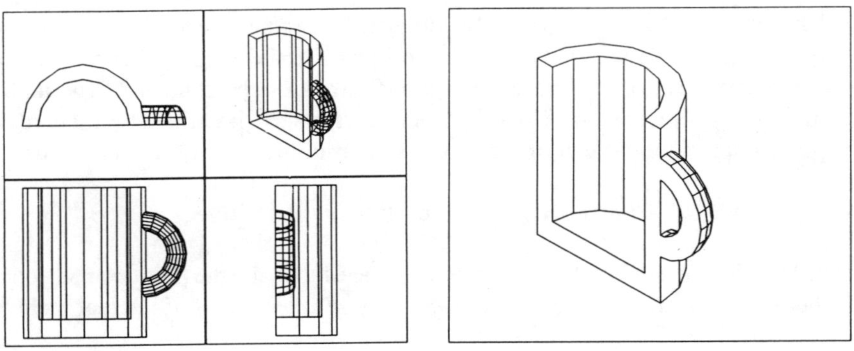

Figure 8.4.2.

8.5 DETAILING ON SELECTED PLANES

C-space command in the **Construct** menu of **SilverScreen** can be effectively employed for adding text lines on different planar surfaces of an object and for dimensioning of its different linear edges, which in general have different orientations. To illustrate such application, Figure 8.5.1 shows a 6′×6′×6′ cube having its six planar

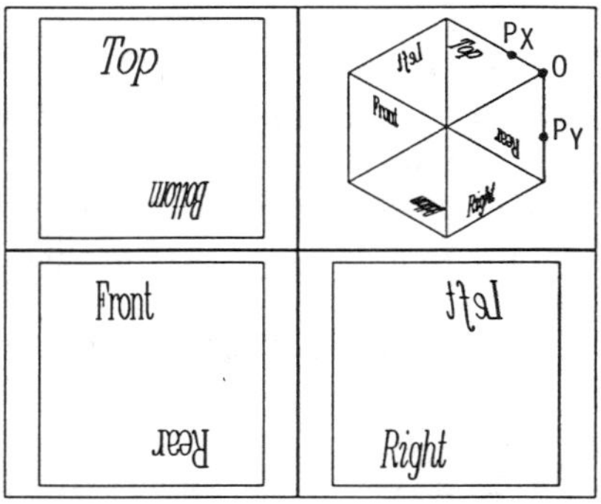

Figure 8.5.1.

surfaces marked with Top, Bottom, Front, Rear, Left, and Right labels. The front surface is easiest to mark. For that, we enter the following commands and responses to the prompted questions:

Text Line
Enter text: Front
Custom Horizontal Left Unslanted
Enter height of character [1]: 1
Enter aspect ratio of character [1.0]: 0.5
Enter spacing ratio between characters [1.0]: 0.25
Select font: roman1
Mark or enter insertion point: 1’6".4’6",6′

The label Front is marked horizontally, left justified, and unslanted.

To label the rear plane, it should first be designated as the working plane by moving the *X*- and *Y*-axes onto that plane. Many options in the **Construct** menu can be applied to achieve this assignment of a C-space. Since the rear plane is defined by a square, we may simply select it by specifying this particular primitive using the commands

Construct C-Space Align Primitive

Arrow keys on the keyboard or a mouse can be used to make the particular selection. Another way is to align *X*- and *Y*-axes along two selected edges of the rear plane by specifying one corner of the rear square to be the origin of the C-space and having any point on a selected edge entered as on the *X*- and *Y*-axis, respectively, such as the points O, P_X, and P_Y in the isometric view shown as the upper, right panel of Figure 8.5.1.

It is good practice to realign a construction space back to the world space. For example, after the label Rear has been added to the rear plane, one is about to mark another plane surface of the cube. Instead of defining a new C-space based on the current C-space, we return to the W-space (*X*- and *Y*-axes on the front plane of the cube) by entering the commands

Construct C-space Align W-space

Figure 8.5.1 has the front, rear, top, and bottom planes all marked using the **roman1** font. The first two labels are unslanted while the last two labels are slanted 15°. The left and right planes are marked using **italic** font and are slanted 15°.

If any entered text line needs modification, the **Text** command in the **Modify** menu can be entered to have a prompt of choices

Aspect Font Move Overscore Rotate
Scale sHear sPacing Underscore

Aspect ratio, spacing, shearing angle, scale, and font style for printing character(s) can all be changed. A character string can also be moved and rotated. And overscore and underscore for the entire string can be added.

The **Construct** menu of **SilverScreen** can also be effectively applied for dimensioning of the linear edges of the cube. Since these linear edges have different orientations, the **Align** command should be applied because it allows the *X*-axis of a C-space to be aligned parallel to a line that needs to be dimensioned. Once a C-space has been defined, the routine detailing work can then be followed. Figure 8.5.2a and b is presented as examples.

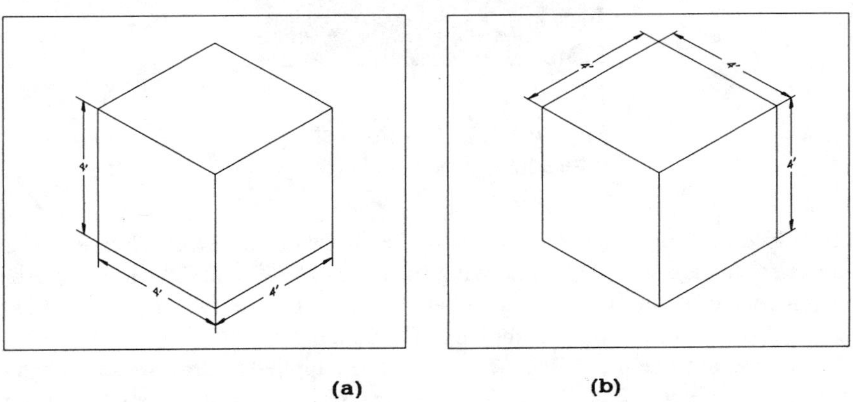

(a) (b)

Figure 8.5.2.

To detail the dimensions of the front plane of the cube, we request the **Annotation** menu and enter the commands **A-block** and **Select**. When the options **Front**, **Right**, and **Top** are shown on screen, we select the **Front** option. The commands **Select**, **Linear**, and **Horizontal** should be entered for dimensioning of the baseline as shown in Figure 8.5.2a. For dimensioning the baseline of the right-side plane of the cube, we have to request the **Annotation** again to change to a new C-space. That is, the commands **A-block** and **Select** need to be entered and then the **Right** option (among Front, Right, and Top being prompted) should be selected as the annotation block. The **Insert** and **Linear** commands can again be entered to dimension the right-side plane; the results of dimensioning the base (using **Horizontal** command) and right (using **Vertical** or **Parallel** command) edges are shown in Figures 8.5.2a and b, respectively.

The dimensionings of the top and left edges of the top plane of the cube shown in Figure 8.5.2b are the results of asking for the **Top** (plane) option as a C-space. More exercises on detailing on selected planes are presented in the next section.

8.6 ISOMETRIC PIPELINE SYSTEMS

Pipeline systems are commonly utilized in the petrochemical industry. Solid modeling of such systems becomes a basic component in many commercial software packages (Figure 8.6.1). **SilverScreen** can be applied to first construct the fundamental components, namely the pipe fittings. Figure 8.6.2 gives a number of the most

Figure 8.6.1. (Courtesy of Algor Interactive Systems.)

commonly used fittings and is drawn with the **Draw** menu of **SilverScreen**. These modules are made available in a drawing file called **PipeFits.DRW**. Making use of them has enabled a three-dimensional pipeline system to be drawn and presented as a four-way, split-screen display as shown in Figure 8.6.3.

The **Construct** menu of **SilverScreen** has been applied to draw these fittings on different planes. For a three-dimensional perspective display of the pipeline system, the isometric view provides a very clear presentation indicating how the straight pipes

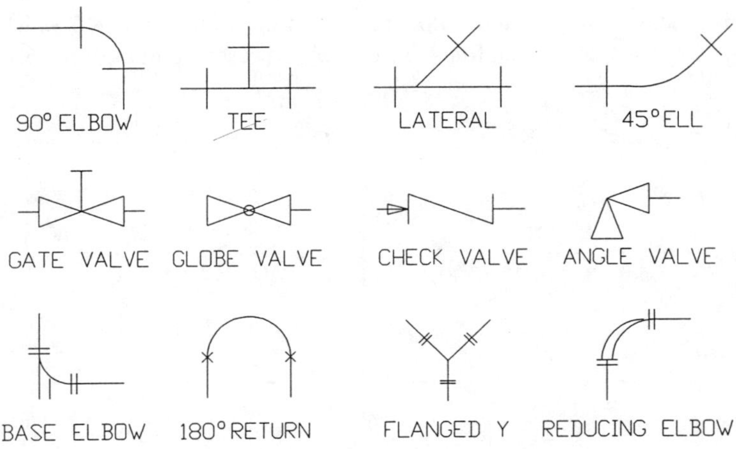

Figure 8.6.2.

are arranged to orient along three orthogonal directions (parallel to the three coordinate axes of the World space).

The dimensions of the pipes in a pipeline system and the various fittings adopted in connecting these pipes can be depicted in a three-dimensional way by application of the **A-block** and **Select** commands in the **Annotation** menu of **SilverScreen**. Figure 8.6.3 illustrates how the **Front**, **Right**, and **Top** options can be selected for labeling the 9′, 5.5′ and 6.5′, and TEE and 90° ELBOW, respectively.

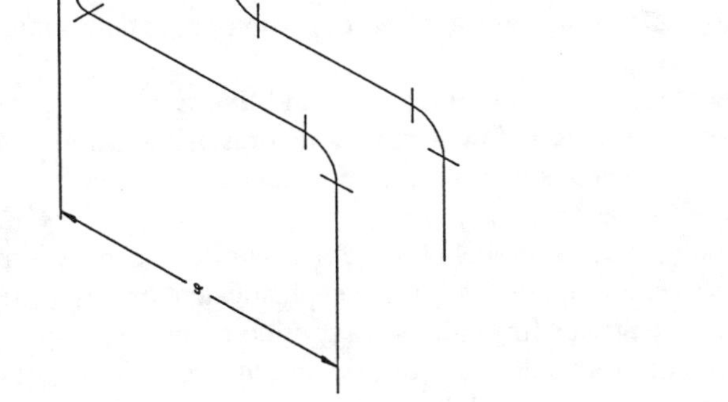

Figure 8.6.3.

To give more details about the pipeline system shown in Figure 8.6.3., which is an isometric view, a split-screen, four-way display is presented in Figure 8.6.4.

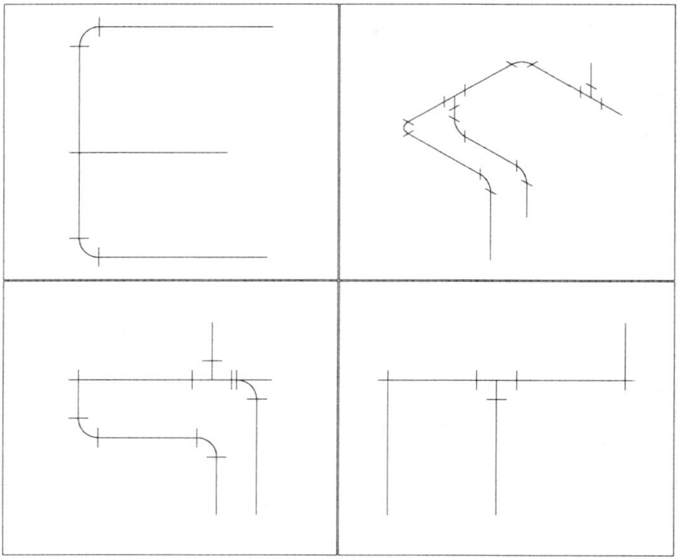

Figure 8.6.4.

8.7 CONCLUDING REMARKS

The sections presented in this chapter demonstrate clearly that proper use of the **C-space** command in conjunction with the commands in the **Annotation** menu of **SilverScreen** facilitates and expedites the construction and detailing of the three-dimensional solid models. Advanced applications of these techniques are possible once the reader becomes more familiar with the procedures involved in the process. Figure 8.7.1 is an example the lastest development in advanced application of the three-dimensional modeling — molecular design and analysis of carbon-based dendrimer for lubricating micron-sized devices.

We could develop a similar program by use of **SilverScreen**. Figure 8.7.2 could be a start-up for the construction of molecular structures using different sizes of spheres with different colors. These spheres can be positioned at different spatial locations by application of the C-space method.

As the three-dimensional models for the real-world objects become increasingly complex, such shown in Figure 8.7.1, more sophisticated displaying techniques need to be employed to clearly distinguish the various components involved in the model, such as assigning different colors and different light intensities to these components, and adjusting the light sources and the viewing positions. In the next chapter, we will discuss such applications so as to make the presentation of the constructed solid models much easier to comprehend.

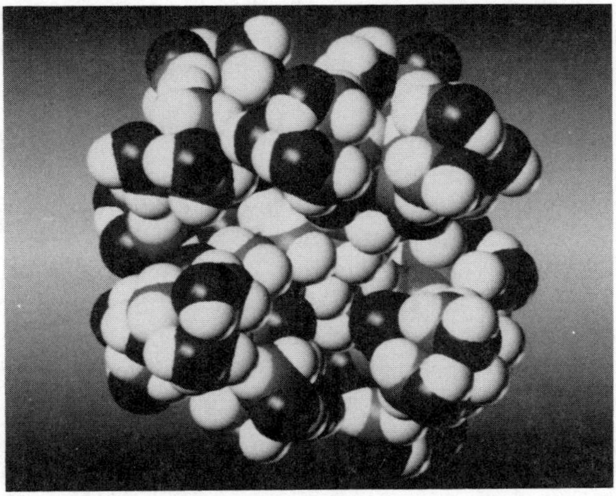

Figure 8.7.1. (Courtesy of Newkome et al., University of South Florida.)

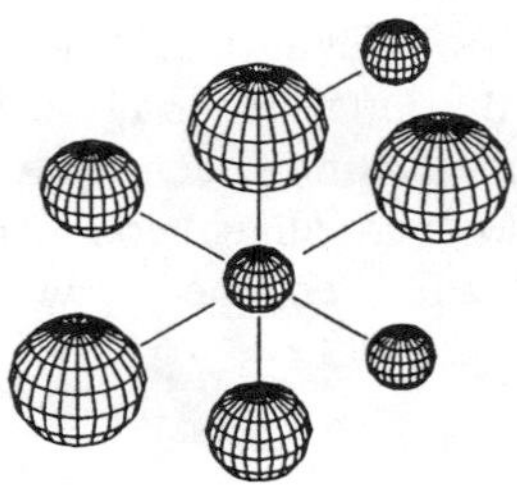

Figure 8.7.2.

PROBLEMS

8.1 First zoom the isometric view of the cube shown in Figure 8.21a and put in the current window, and then add labels A through H near the vertices of the cube.

8.2 Add a string of characters "Inclined Plane" on the oblique plane ABCD shown in the isometric view of Figure 8.2.1b. Try to place the string of characters parallel to the longer side of the plane either in a single line or two lines (one word per line).

8.3 Place the isometric view shown in Figure 8.2.4 in the current window and then add the three-dimensional detailing of the object including the sides and also the cylindrical hole.

8.4 Divide the coffee mug shown in Figure 8.4.1 into two halves by using a cutting plane that is parallel to the bottom plane of the mug and passes through the exact middle of the handle. Present a four-way, split-screen display of the lower half, having all of the hidden lines removed in the isometric view.

8.5 The labels Top and Bottom on the cube shown in Figure 8.5.1 are slanted at an angle of 15°. Apply the command **sHear** so that they become unslanted.

8.6 Change the labels Front and Rear on the cube shown in Figure 8.5.1 to the italic style.

8.7 Apply the **Rotate** command so that the label Right on the cube shown in Figure 8.5.1 is turned upside down.

8.8 Label all of the pipe fittings appearing in Figure 8.6.3 by application of the **A-block** command of the **Annotation** menu.

8.9 Detail the dimensions of all of the pipes appearing in Figure 8.6.3.

8.10 Generate a sphere of any size by application of the **Sweep** command in the **Geometry** menu. This sphere is then to be copied into two other spheres of different sizes. Finally, apply these three spheres of different sizes for construction of a molecular model as shown in Figure 8.6.2. The **C-space** command in the **Construct** menu should be effectively used in this solid modeling process.

8.11 Place the characters A, B, C, D, E, and F, one each, on the six planar surfaces of a cube. Save the resulting drawing in a drawing file called **Cube6Cha.DRW**. Try different available fonts of **SilverScreen** for these six characters.

8.12 Generate a pipeline system similar to that shown in Figure 8.6.4 and then save the drawing in the file **Paof864.DRW**.

8.13 Load the drawing file **Paof864.DRW**, detail its isometric view to produce a drawing similar to that shown in Figure 8.6.3, and save it in the file **Paof863.DRW**.

8.14 Some may complain that the dimensions and labels in Figure 8.6.3 are too small. The **Zoom** command in the **Window** menu of **SilverScreen** can be effectively applied to enlarge any area of particular interest. Apply this method to the drawing file **Paof863. DRW** to obtain an enlarged view as illustrated in Figure P8.14.

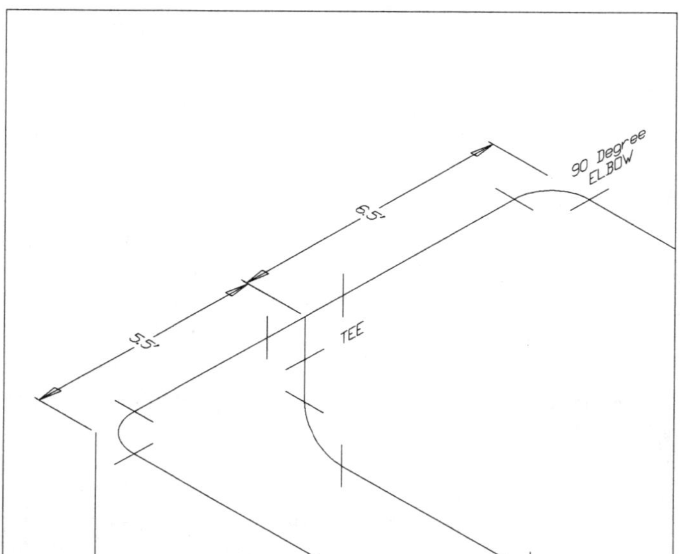

Figure P8.14.

8.15 Generate a cube and apply the **Annotation** menu of **SilverScreen** to add labels by using the **Insert** and **Note** commands. **A-block** and **Select** commands are to be used to select the front, right, and top surfaces of the cube shown in Figure P8.15.

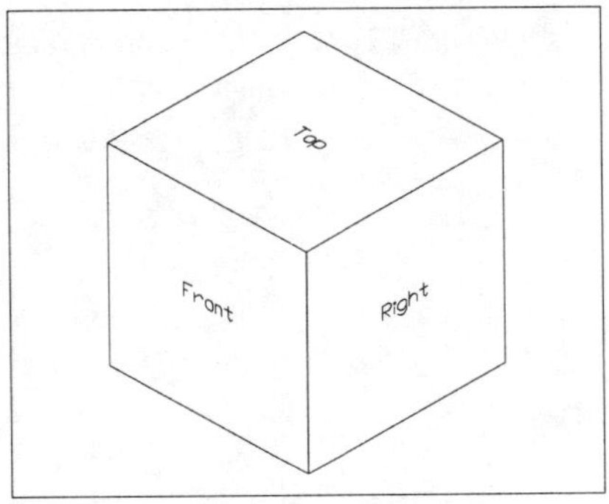

Figure P8.15.

8.16 Generate a hollow cylinder and add a label such as "Y. C. Pao's mug" shown in Figure P8.16 on one of its outside, lateral planes. To select one such plane, the **A-block**, **Create**, and **Align** commands should be effectively applied. By interactively selecting an outside polygon of the cylinder and giving it a name, for example **plane**, it will be added to the original selection list of Front, Right, and Top. By entering the **Select** command and picking the **plane** option, a desired label can be placed using the **Insert** and **Note** commands.

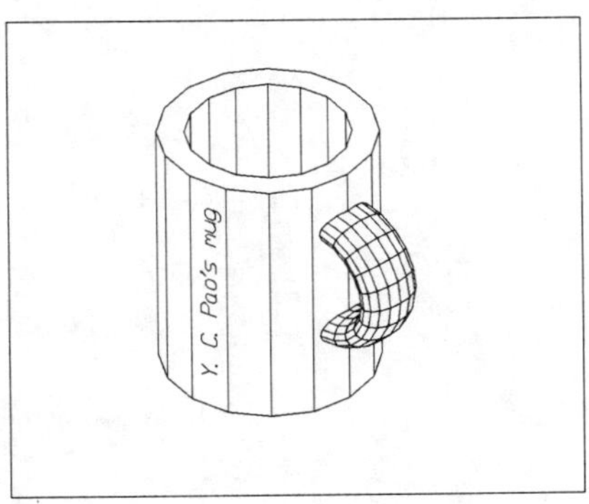

Figure P8.16.

9 Presentation of Constructed Objects

9.1 INTRODUCTION

In solid modeling, the ambiguity such as in the case shown in Figure 1.2.1a, where it is difficult to discern whether the circular disk is an insert or a cutout, may not be totally eliminated by the application of the hidden-line removal technique alone. It can be further eliminated by application of different degrees of *shading* to its surfaces and by placing the *light source* at a well-selected location. A pacifier is shown in Figure 9.1.1, which illustrates how it will be illuminated under a certain light source and where the shadow will be cast. **SilverScreen** has commands that can be effectively employed to change the location of a light source and also to select different shade and shadow patterns to paint the surfaces of a three-dimensional solid model.

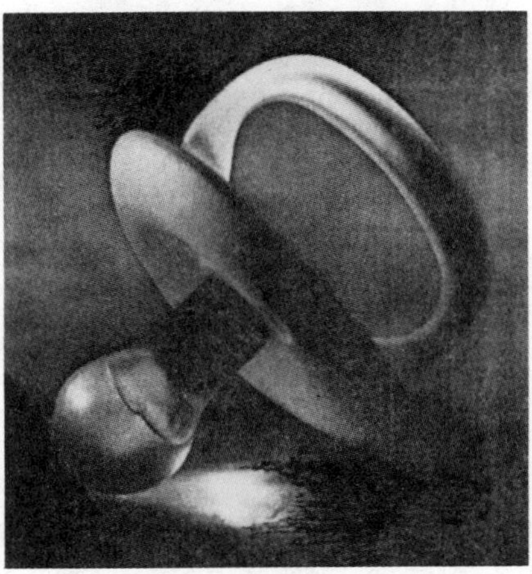

Figure 9.1.1.

If a color monitor is available to **render** the display of a solid model, it also provides a powerful tool for visual validation whether or not the model truly represents a three-dimensional object. A solid model of a bolt shown in Figure 9.1.2 clearly indicates the advantage of having a color display monitor in visualization of the three-dimensional object. This type of color-filling technique can also be applied effectively for **contour**

plotting of the computed results such as a stress or temperature distribution in a three-dimensional object (Figure 9.1.3). **SilverScreen** has color-selection and color-filling commands for implementing such rendering and contour-plotting tasks.

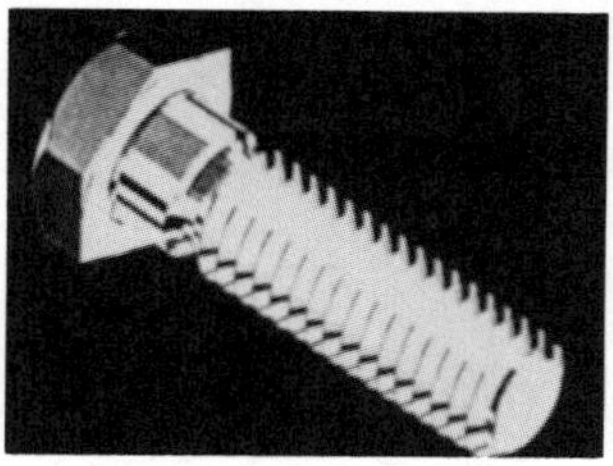

Figure 9.1.2.

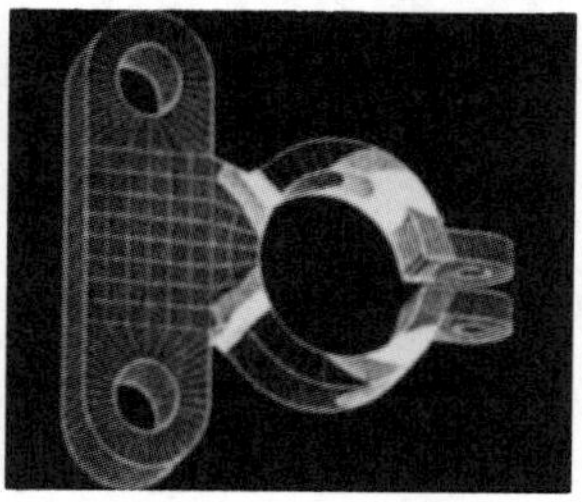

Figure 9.1.3.

(Courtesy of Algor Interactive Systems, Pittsburg, PA.)

In this chapter, we will introduce methods for shading the surfaces of a solid model when using a black-and-white or monochrome display monitor. Different **patterns** provided in **SilverScreen**'s file directories will be selected to fill the surfaces of a solid model. Also, by use of a color monitor, the *color selector* provided by **SilverScreen** will be introduced to enable the users to assign different colors for different surfaces of a solid model or to assign different colors for different objects for the purpose of distinction. Objects can also be shown with **edges** or without edges even though their solid models are all constructed with edges. Various methods of presentation are discussed in this chapter for either monochrome or color display of solid models.

9.2 COLOR FILLS

In solid modeling, polygons are the most commonly applied primitives for constructing the external boundaries of an object. To differentiate the different polygons involved in a solid model from each other, they may be shaded differently or filled with different colors or patterns. In application of **SilverScreen** for solid modeling, all polygons are filled with **white** color unless otherwise arranged. In this section, we discuss methods of applying the **SilverScreen** comands to fill such polygons with different, specified colors. In the next section, we will introduce available patterns that can also be applied for filling the polygons to achieve the same objective of identifying individual surfaces.

Let us first discuss how to fill a polygon with a designated color. A cube with six faces is certainly a good object to illustrate such a color filling process. We may extract a cube from the drawing file **PaoSolid.DRW** with the commands **I/o** and **Xcopy** or by simply constructing it anew by a sweeping process with the commands **Draw**, **Polygon**, **Rectangle**, **Geometry**, **Sweep**, and **Linear**. After the cube has been created or copied, it is better to have an enlarged isometric view in the current window so that its six faces can be viewed clearly and all easily identifiable as possible primitive candidates for color filling. This is done by first using the commands **Window** and **eXchange** to interchange the current window (lower, left panel of the graphics area) showing the front view with the window showing the isometric view (upper, right panel of the graphics area). The enlargement of the isometric view can be achieved by entering the commands **Window** and **Enlarge**. A newer version of **SilverScreen** allows a window to be enlarged by entering the commands **Window** and **Enlarge** directly, without going through the **eXchange** command, by using the arrow keys on the keyboard or the pointing device to select the desired window.

Let us fill the front, back, right-side, left-side, top, and bottom faces of the cube with colors having code numbers 1 through 6 listed in the **color selector** (Figure 9.2.1[1]), respectively. To fill the front face with color #1 (blue), we enter the commands

Presentation Surface Primitive (1)
(Use arrow keys or the pointing device
to select the front face.)
Fill-color
Select or enter new surface color: 1

No change will take place in the graphics area of the screen when the fill-color command has been completed. The only indication is that **Silver** menu reappears on screen. To determine whether the front face of the cube has in fact been filled with the blue color, we have to enter the commands

Presentation Hide Object (2)
(Press the Enter key because the cube is
the only object.)
Quick Edged
Select or enter color for edges: 15 (white)

By following procedure (1), we can also use arrow keys or the pointing device to select the back, right-side, left-side, top, and bottom faces of the cube, and request colors numbers 2 through 6 from the color selector to fill their respective faces. After all six faces have been filled, procedure (2) can be followed to obtain a hidden-line-removed display of the cube as shown in Figure 9.2.2.[1] Since it shows only the front, right-side, and top faces of the cube, a rotation about the X-axis by 180° followed by another rotation about the Y-axis by –90° has made it possible to produce another color display, Figure 9.2.3,[1] of the cube showing its left-side, back, and bottom faces.

In case no color plotter and no color printer is available, a hard copy using a black-and-white laserjet printer can be obtained by following the procedure previously mentioned in Section 5.2. The colors listed in Figure 9.2.1 when printed through a

---

[1] See outside back cover.

black-and-white laserjet printer will appear in different *grayscales* as shown in Figure 9.2.4. In grayscales, the hard copies for replacing the two color displays shown in Figures 9.2.2 and 9.2.3 will appear as in Figures 9.2.5 and 9.2.6.

As an application of the color-filling technique, the cross section at which the mug shown in Figure 8.4.2 is cut into two halves can be highlighted with a fill of the red color while the all edge lines can be colored with the blue color. The resulting grayscale hard copy is shown in Figure 9.2.7.

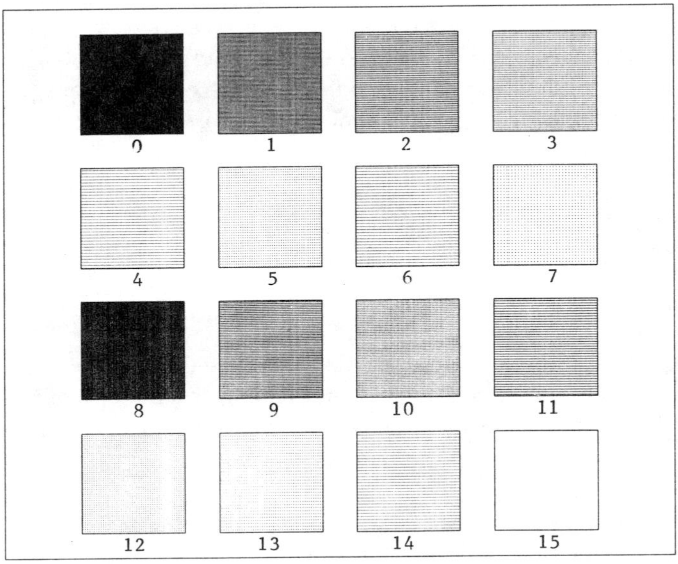

Figure 9.2.4.

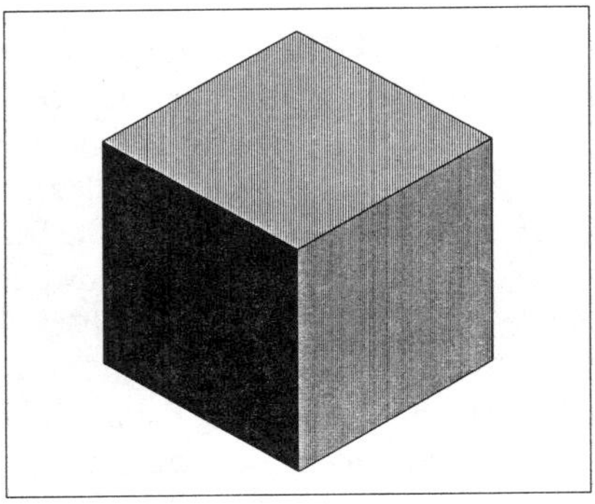

Figure 9.2.5.

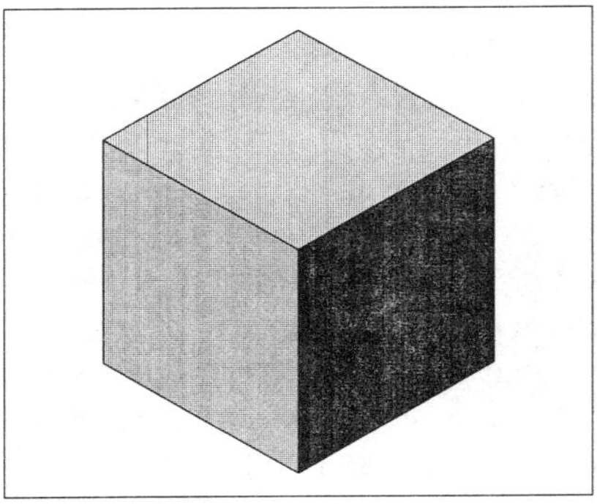

Figure 9.2.6.

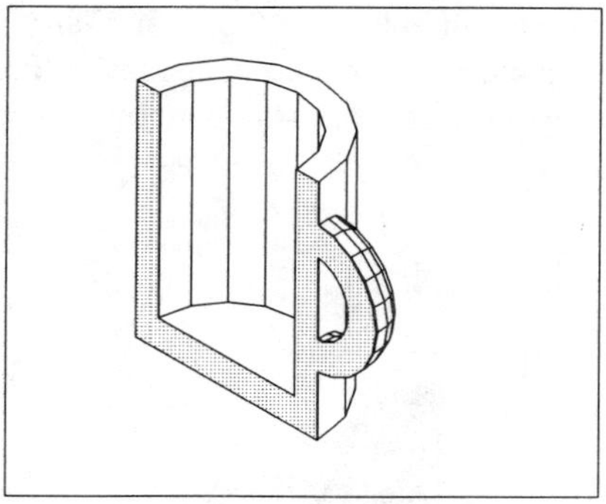

Figure 9.2.7.

9.3 PATTERN FILLS

Instead of filling the polygon surfaces of a solid model with a specified color, any pattern such as those standard fill patterns listed in Appendix A can be selected to achieve the same purpose of identifying individual polygons. However, it should be particularly pointed out that the selected pattern will be drawn in **white** color unless prior arangement has been made to change the pattern color. Since all polygons will initially have a white background (unless it has been later altered), the pattern lines also drawn in white will be invisible! Hence, to fill any polygon with a selected pattern, its background color must first be changed to any color other than white.

To demonstrate the process of pattern filling, let us again adopt a cube and have its front, right-side, and top faces filled with the **brass**, **steel**, and **star** patterns, respectively. To fill the front face of the cube with the **brass** pattern, we enter the commands

Presentation Surface Primitive
(Use the arrow keys on keyboard or the pointing device to select the front face.)
Pattern
(Use the arrow keys on keyboard or the pointing device to select the **brass** pattern from among all of the available patterns prompted on screen.)

After completion of the last command, no pattern filling of the front face of the cube will actually be displayed on screen; only the **Silver** menu will be redisplayed. For an inspection of the one-face-filled cube, again we have to use procedure (2) presented in Section 9.2. The resulting display of the cube having its front face filled with the brass pattern is presented in Figure 9.3.1.

By following the procedure just described, the right side of the cube can also be filled with the **steel** pattern. The display of the cube having two of its six faces filled with the patterns **brass** and **steel** is shown in Figures 9.3.2. Similarly, the top face of the cube can be filled with the **star** pattern. The resulting display of the cube having three of its six faces filled with the **brass**, **steel**, and **star** patterns, respectively, is presented in

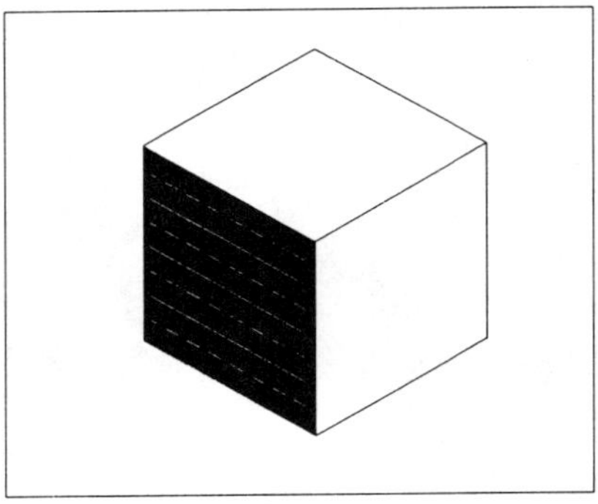

Figure 9.3.1. The front face is filled with the **brass** pattern.

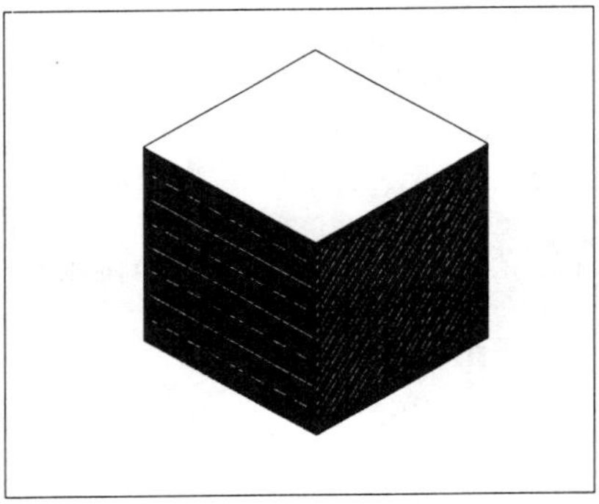

Figure 9.3.2. The front and right-side faces are filled with the **brass** and **steel** patterns, respectively.

Figure 9.3.3. After the back, left-side, and bottom faces of the cube filled with patterns **brick**, **dots**, and **hex**, respectively, we can display these three faces by a rotation about the *X*-axis (W-space) by 180° (using the commands **Rotate** and entering x180) then followed by a rotation about the *Y*-axis (W-space) by 180° (using the commands **A-rotate** and entering 180) and obtain a hard copy as presented in Figure 9.3.4.

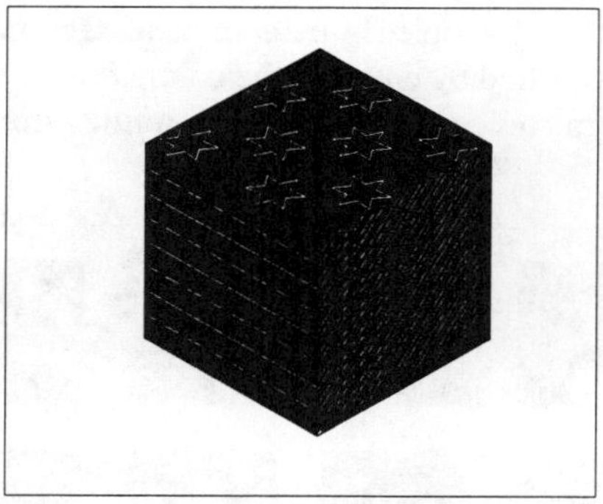

Figure 9.3.3. The front, right-side, and top faces are filled with the **brass**, **steel**, and **star** patterns, respectively.

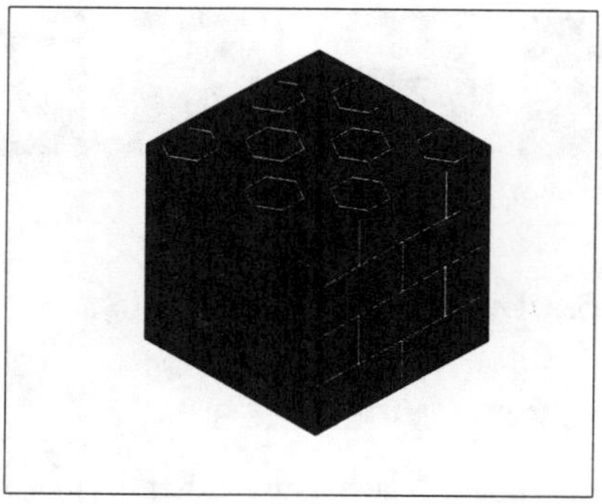

Figure 9.3.4. The other three faces of the cube are filled with the **brick**, **dots**, and **hex** patterns.

All faces of the cube shown in Figures 9.3.1 through 9.3.4 have been filled with a black ground so that the pattern lines drawn in white color are easily visible. We will later show how the color for drawing the pattern lines can be specified in any available color other than white. It will then be possible to draw the pattern lines on a white background.

As another example of the pattern-filling exercise, Figure 9.3.5 is presented to show 16 squares arranged in a matrix form of four rows and four columns and filled with the first 16 patterns listed in Appendix A. In addition, the name of each pattern is labeled at the bottom of each square by application of the **Text** command of **SilverScreen**. One square is first drawn and then it is copied by the commands **Geometry**, **Copy**, and **Matrix**. Each square is next color filled by entering the color code "0" and pattern filled with one of the first 16 pattern choices in Appendix A, **angle** through **dash**.

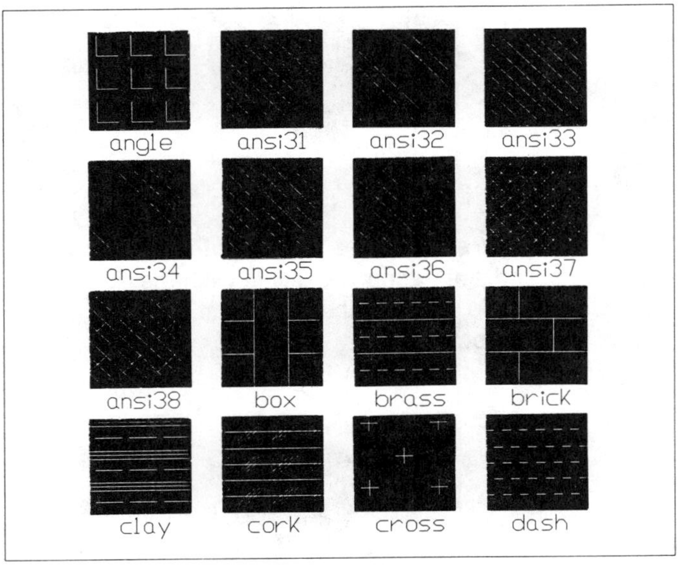

Figure 9.3.5. Display of 16 fill patterns.

9.4 PRESENTATION WITH WIDE LINES

In solid modeling, objects can be drawn using either normal-size or **wide** lines. Such a choice can be made by use of the **Force** command in the **Presentation** menu of **SilverScreen**. In Figure 9.4.1, a circle drawn with normal-size lines and a square drawn with wide lines are shown. A script file called **PaoWline.SF**, which can be executed using the commands **Language**, **Execute**, and **Continuous**, is listed below to give the steps involved in generating this display.

Script File **PaoWline.SF**:

```
create object circle at 9,9,0
draw circle radius 3.5,0,0
zoom drawing
force enable object\circle width wide
create object square
draw rectangle 8,10,0 10,8,0
goto object \circle
force enable object \circle width normal
```

goto object \square
force enable object \square width **wide**

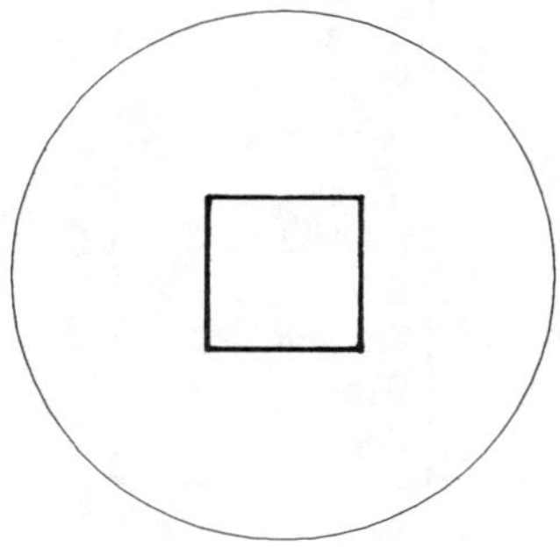

Figure 9.4.1.

The screen will show that the circle is first drawn with normal-size lines and then changed to using the wide lines. The square is next drawn with normal-size lines. The **Force** command is next applied to interchange the line size, using wide lines for the square and normal-size lines for the circle. The procedure for applying the **Force** command is

Presentation Force Enable Object Width Normal/Wide

The **Create** and **gOto** commands in the **Structure** menu of **SilverScreen** can be effectively employed to switch from one object to the other. This is demonstrated in the program **PaoWline.SF** when dealing with two objects **Circle** and **Square**.

9.5 CAMERA WORK

Just like taking a picture, the command **Camera** in the **Window** menu (Figure 9.5.1) of **SilverScreen** can be applied obtain a **snapshot** of the object being displayed on screen. This snapshot can be taken at any angle just as when taking an auxiliary view. A snapshot file with an extension **.PIX** will be created by **SilverScreen** and stored in the **DRW** subdirectory.

For example, we may retrieve the drawing file **Paof823.DRW** to have Figure 8.2.3 displayed or reedited on screen. To take a snapshot of this four-way, split-screen display, we enter

Window Camera Snapshot Take
Enter snapshot name: **Paof823**

A file **Paof823.PIX** will be created and stored in the **DRW** subdirectory of **SilverScreen**. To view a snapshot, we need to know its **.PIX** file. In case we wish to view the file that

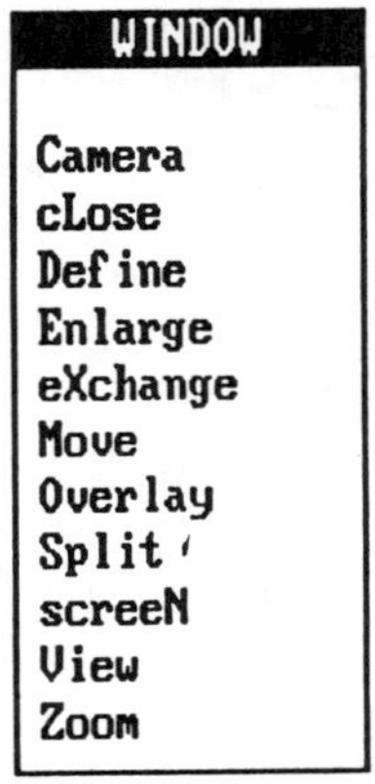

Figure 9.5.1.

has just been created, **Paof823.PIX**, the commands to be entered are

Window Camera Snapshot View
Enter snapshot name: **Paof823**

Notice that there is no need to mention the extension **.PIX**. Although the last question about the file name is prompted on the top line of the screen, a list of all **.PIX** files currently stored in the **DRW** subdirectory of **SilverScreen** will appear on screen. The user has the choice of either using the keyboard to type in the file name or using a pointing device to select the desired file from the displayed list.

9.6 LIGHT SOURCE AND SHADING

The positions of a group of three-dimensional objects relative to each other can be clarified by displaying them on a two-dimensional screen by effective shading of the surfaces of the objects in reference to a specified **light source**. To illustrate this concept, let us arrange to have two objects, a block of size 4 by 4 by 6.25 and a pyramid having a base of size 4 by 4 and a height of 4, resting on a foundation that has an area of 20 by 20, all in generic units. And we also specify a light source so that the surfaces of these objects can be properly shaded. These objects are shown in Figure 9.6.1 along with the coordinate system used and the location of the light source (20,12,–10). In this section, we introduce the **SilverScreen** commands that will enable this assembly of three-dimensional objects to be shaded and displayed on screen.

A hidden-lines-removed view of the arrangement is shown in Figure 9.6.2. The objects are clearly marked with BLOCK, PYRAMID, and FOUNDATION by application of the **C-space** command in the **Construct** menu and the **A-block** command in the **Annotation** menu of **SilverScreen**. The light source (using slanted characters) and the World space are also properly marked. Figure 9.6.2 is the result display before the shading is applied.

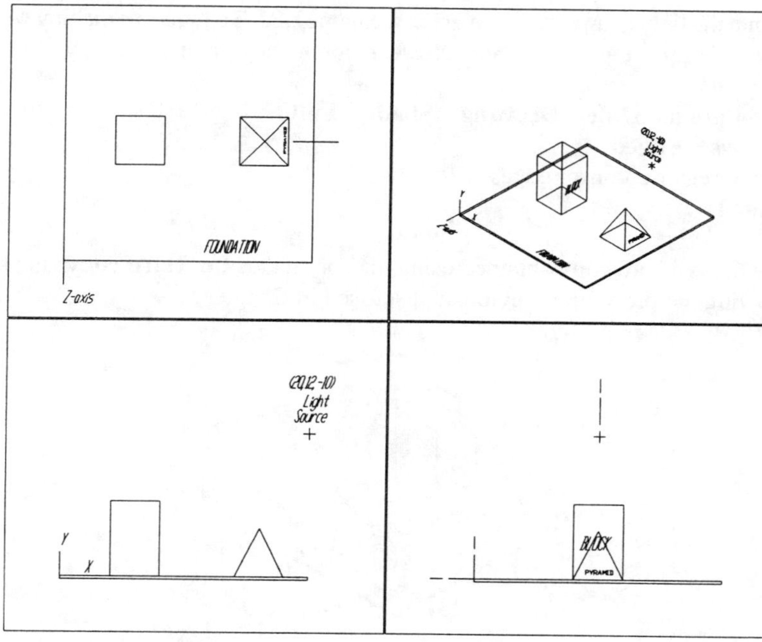

Figure 9.6.1.

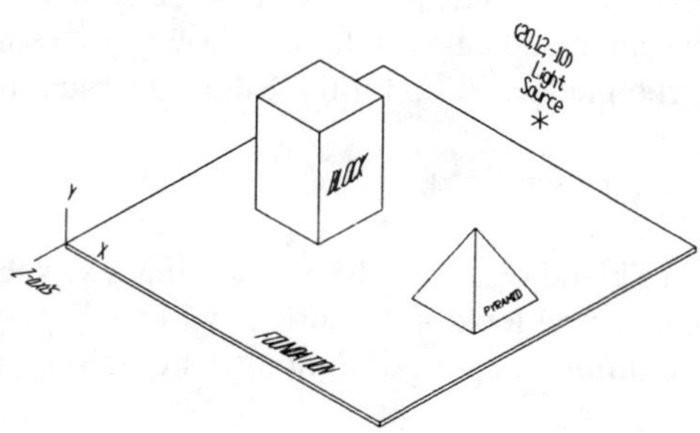

Figure 9.6.2.

To define a light source, the **SilverScreen** commands to be entered are

Presentation Light Source Define
Enter light name: **LS1**
Enter or mark location of light source: 20,12,–10
Enter contrast (1-100): 50
Enter RGB mixture (0-255 in format n,n,n): 50,50,50
(Equal values for having "white" light.)

Notice that the light source has been given the name **LS1**. To request a display with the **shading** technique applied, the **SilverScreen** commands involved are

Presentation Hide Drawing Shade Full
Shadows Edged
Select or enter color for edges: 12
Blended

Figure 9.6.3 is a hard copy obtained using the commands **I/o**, **Hard copy**, and so on, by following the procedure previously discussed in Chapter 5.

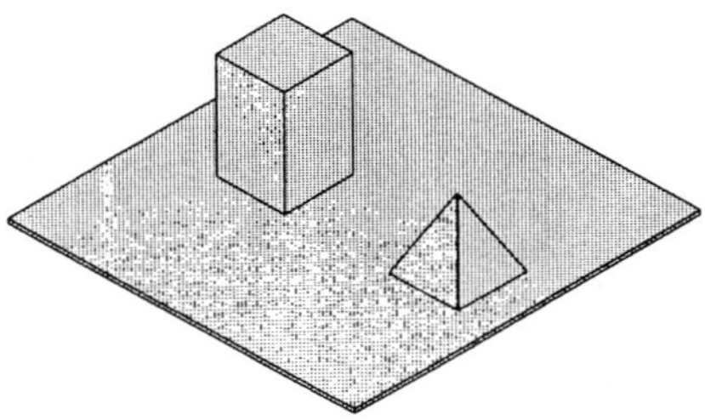

Figure 9.6.3.

If a surface has not been filled with a selected color, a default color number 15 will be assigned. The **Edge** command enables the visible edge of a surface to be assigned a desired color. The prompted question for the **Edge** command is

Select or enter color for edges:

The user may choose to blend or unblend for rendering a surface. When the **Blend** command is selected, a natural look of the surface will be displayed by reducing the sharp contrast between different levels of light intensity falling on the surface.

PROBLEMS

9.1 Select one among the objects shown in Figure P9.1 to use the wide-line option of **SilverScreen** to draw it.

9.2 Apply the **Construct** and **Annotation** menus to change the labels of the light source and the coordinate axes shown in Figure 9.6.2 so that they will appear as in Figure P9.2.

9.3 Define a new light source at (0,12,0) and use a new name **LS2** to produce a new shaded view and then to make a hard copy of the block and pyramid placed on the foundation shown in Figure P9.2.

9.4 Reinitiate the construction of the cube whose six surfaces are filled with the different fill patterns shown in Figures 9.3.4 and 9.3.5 by creating a script file

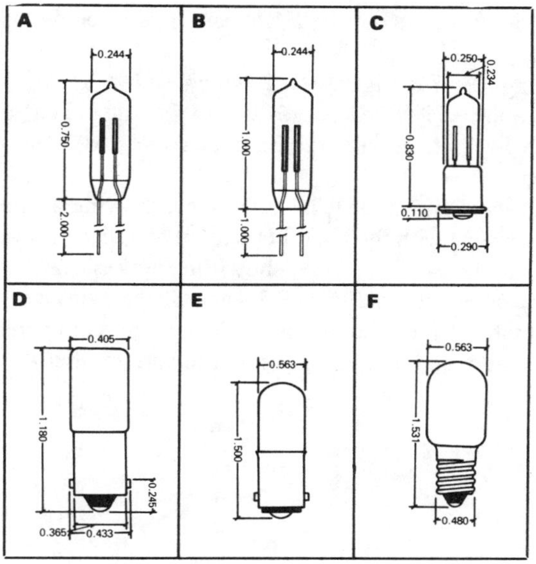

Figure P9.1.

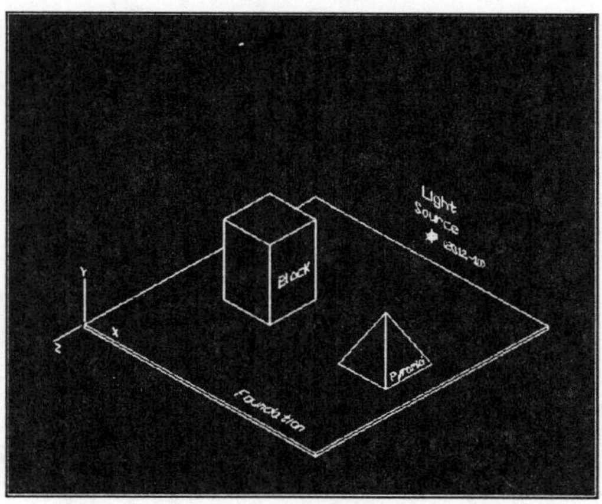

Figure P9.2.

called **Paof9345.SF**. Refer to Chapter 5 for the application of the command **Remember** in the **Language** menu of **SilverScreen**.

9.5 Apply the **Text-edit** command in the **Language** menu of **SilverScreen** for the script file **Paof9345.SF** created in Problem 9.4 to replace the **STARS** pattern with the **ZIGZAG** pattern. See Appendix A for the available **SilverScreen** fill patterns.

9.6 Same as Problem 9.5 except the **HEX** pattern is to be replaced by the **NET3** pattern.

9.7 If the drawing file for Figure 9.4.2 which shows an assembly of 8 three-dimensional objects is named **Paof542**, load this file to generate a snapshot file **Paof542.PIX** by application of the **Camera** command in the **Window** menu of **SilverScreen**.

9.8 The half-mug shown in Figure 9.2.7 is saved in a drawing file **Paof972.DRW**; fill its top surface (rim) with the **EARTH** pattern and then produce a hard copy.

9.9 Paint the six surfaces of the cube shown in Figures 9.2.5 and 9.2.6 with six different grayscales listed in Figure 9.2.4 and label the surface with the respective grayscale number by effective application of the command **A-block** in the **Construct** menu and the **C-space** command in the **Annotate** menu of **SilverScreen**.

10 SilverSmith: Combining SilverScreen with BASIC and C Programs

10.1 INTRODUCTION

This chapter introduce the compiling and execution capabilities, called **SilverSmith**, of **SilverScreen** for dealing with programs written in **BASIC** and **C** languages. It allows **SilverScreen** commands to be combined with the statements in **BASIC** or **C** language to develop **script files** that can be run by the **Execute** command of **SilverScreen**. This feature is important for the **SilverScreen** users who already have a large number of **BASIC** and/or **C** programs developed for calculations and generation of plots, and are anxious to execute these programs in conjunction with **SilverScreen**.

Any software for the word-processing needs, such as WordPerfect[1] or VEDIT,[2] and others can be used for editing a **SilverSmith** program developed with either **BASIC** or **C** language. The statements involving the **SilverScreen** commands appearing in such programs must start with a character **@**. And there are other rules to be discussed in this chapter that must be followed in developing the **SilverSmith** programs. The edited **BASIC** and **C** programs are to be transferred into the **EX** subdirectory of **SilverScreen** in the form of **ASCII** (American Standard for Information Interchange) files with an extension of **.BAS** or **.C**.

The **Language** menu of **SilverScreen** (Figure 10.1.1) is to be employed for compiling a BASIC or C program by entering the respective commands of

Language Compile BASIC filename

and

Language Compile C filename

SilverScreen also has the **Text-edit** command in the **Utility** menu (Figure 10.1.2) that can be applied for editing if a **BASIC** or **C** program needs to be changed during

---

[1]WordPerfect is developed by Satellite Software International, Orem, Utah.
[2]VEDIT is developed by CompuView Products, Inc., Ann Arbor, Michigan.

compilation. Successful compilation will create an executable file with an extension **.EX** and this file can then be run and viewed on screen by entering the commands

Language Execute filename

```
LANGUAGE

Compile
cUstom-menu
Execute
Macro
Prompt
Remember
Text-edit
Write
```

Figure 10.1.1.

```
UTILITY

Clear
Execute
Help
Linestyle
Multi-window
Penstyle
Startup
Text-edit
Undo
uNits
```

Figure 10.1.2.

In the following sections, a series of examples will be presented to help explain how **SilverSmith** programs can be developed, compiled, and executed.

10.2 APPLYING SILVERSCREEN COMMANDS IN BASIC PROGRAMS

As a first example of applying the **SilverScreen** commands for drawing a simple straight line from (0,0,0) to (5,5,0), one may use the **Draw** and **Line** commands along with the **BASIC**[3] statements as follows:

```
'Setup variable as SOME
DIM SOME AS INTEGER
SOME = 5
@ Draw Line 0,0,0 {SOME},{SOME},0
```

This first statement starting with a single quote or apostrophe is a *comment* statement. The second statement declares SOME as an integer variable. Then this variable SOME is assigned to have a value equal to 5 in the third statement. The fourth statement starting with a character **@** signals a **SilverScreen** statement—in this case for drawing a line from the beginning point whose three-dimensional coordinates are (0,0,0) to the ending point whose three-dimensional coordinates are (SOME,SOME,0), having the value of SOME already defined. The curly braces, **{}**, are required by **SilverSmith** to execute the command at the run time.

[3]Any BASIC software such as QuickBASIC developed by Microsoft Corporation, Redmond, Washington.

To save a **SilverSmith** program written in **BASIC** language and display the resulting drawing on screen by application of **SilverScreen**, additional steps need to be taken. Let us demonstrate the complete procedure by drawing a circle with a **BASIC** program in combination with some **SilverScreen** commands. A program called **SSmithC.BAS** developed for that purpose is listed below.

```
' Program SSmithC.BAS — SilverSmith Circle.Basic
'          Draws a circle of radius r and center
'          at (xc,yc).
    r = 20.0
   xc = 0.0
   yc = 0.0
   xb = r+xc
   yb = yc
 for i = 30 to 360 step 30
   ar = i*3.14159/180.0
   xe = xc+r*cos(ar)
   ye = yc+r*sin(ar)
 @draw line {xb},{yb},0.0 {xe},{ye},0.0
 xb = xe
 yb = ye
 next i
```

The above program can be edited by use of any available word-processing software package but should be saved as a **DOS ASCII** file. In order for **SilverScreen** to compile and execute, this file should be copied into the subdirectory **EX** by a DOS statement

A>copy SSmithC.BAS C:\Silver\Ex

It is assumed that the program is stored on a disk presently resided in the drive A and the **SilverScreen** system is in the drive C.

Next, the program **SSmithC.BAS** needs to be compiled, the **SilverScreen** menu and commands required are

Language Compile Basic SSmithC

This creates an executable file **SSmithC.EX** which can be run by pressing the **F8** key and then entering the file name **SSmithC**, or using a sequence of commands such as

Language Execute filename (SSmithC)
Coninuous or **Single-Step**

The program can be run either continuously or in a step-by-step manner by selecting the **Continuous** or **Single-Step** command. If the scale of the object is different from that of the current window, the drawn graph may be out of range and a request for

Window adjustment such as using the **Zoom** or **Four-Way** command may be necessary for displaying it properly on screen.

Figure 10.2.1 is a four-way, split-window display of the result when the program **SSmithC.BAS** is run with **SilverScreen**. The circle has a center at (0,0,0) and a radius equal to 20, and is drawn using an angle increment of 30°. For such a large angle increment, the shape is a polygon instead of the expected circle. Toward achieving the objective of drawing a smooth circle, a problem is given at end of this chapter for readers to modify the program **SSmithC.BAS**.

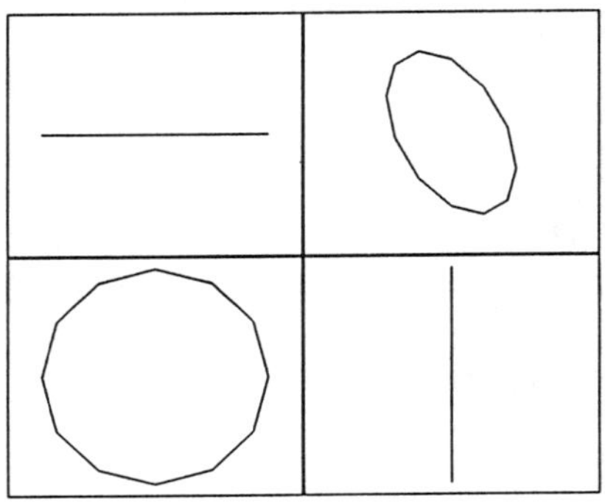

Figure 10.2.1.

10.3 MORE EXAMPLES OF SILVERSMITH–BASIC PROGRAMS

Suppose that one wishes to draw a rectangular gridwork as shown in Figure 10.3.1. It is on the $Z = 0$ plane and the vertices have X coordinates ranging from 1 to 8 and Y ranging from 3 to 7, all in generic units. The increments used for the X and Y coordinates are 1 and 0.5, respectively. Program **SSmithRG.BAS** listed below is developed for the purpose of drawing this rectangle.

```
' Program SSmithRG.Bas — SilverSmith Rectangular Grid in Basic
'         Draws a rectangular grid from x1 to x2 and y1 to y2
'         using increments xi and yi, respectively.
    x1 = 1.0 : x2 = 8.0 :          y1 = 3.0 :          y2 = 7.0
    xi = 1.0 : yi = 0.5
 for x = x1 to x2 step xi
   @ draw line {x},{y1},0.0 {x},{y2},0.0
   next x
 for y = y1 to y2 step yi
   @ draw line {x1},{y},0.0 {x2},{y},0.0
   next y
```

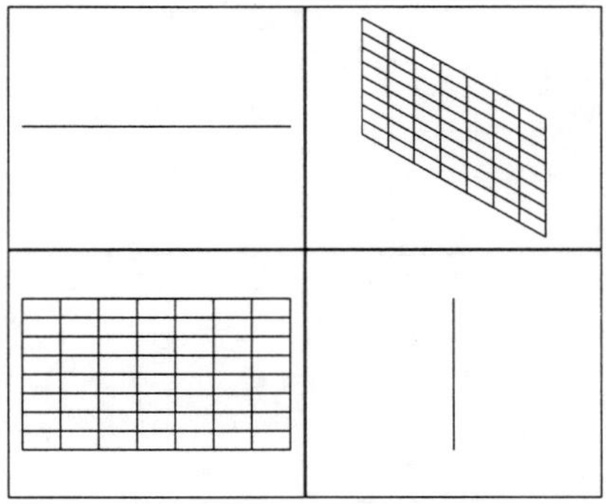

Figure 10.3.1.

In Section 2.8, we applied the **F-Mesh** command of **SilverScreen** for drawing an involute profile of a gear tooth. Here, let us attempt to do it by a **SilverSmith** approach. Program **SSmithIv.BAS** listed below is written in **BASIC** language aided by some **SilverScreen** commands. The resulting four-way, split-screen display of the involute profile is shown in Figure 10.3.2. This profile is drawn on the $Z=0$ plane and on a base circle whose center is at (0,0) and with a radius equal to 200. The unwrapping is started from an angle equal to 95° and ended at an angle equal to 60°. The profile consists of 10 linear segments (11 points) generated with an angle decrement of 3.5°.

```
'Program SSmithIv.BAS — Draws an involute with SilverScreen
'                        and BASIC
      a1  = 95.0        : a2 = 60.0     : da  = 3.5     : pi = 3.14159
    conv  = pi/180.0    : rb = 200.0    : a   = a1      : t  = a*conv
      xb  = rb*cos(t)                   : yb = rb*sin(t)
FOR i = 2.0 TO 11.0
      a = (a1-(i-1.0)*da)*conv : rt = rb*(i-1.0)*da*conv
      x = rb*cos(a)-rt*cos(a-pi/2.0)
     ye = rb*sin(a)-rt*sin(a-pi/2.0)
@draw line {xb},{yb},0.0 {xe},{ye},0.0
     xb = xe : yb = ye
   NEXT i : END
```

To actually draw a gear tooth using involute profiles, we expand the program **SSmithIV** to generate a new program called **SSmithTP.BAS** based on the symmetry property. As shown in Figure 10.3.3, which is a four-way, split-screen display of the gear tooth obtained by using **SSmithTP.BAS** and **SilverScreen**, we arrive at the desired tooth profile by taking a mirror image of the single involute profile obtained in Figure 10.3.2 relative to the *Y–Z* plane and by connecting the two top points. In the program **SSmithTP**, the first and second **@**draw line statements draw the left and right (mirror image) involute profiles, respectively, while the third, **@**draw line statement connects the two top points.

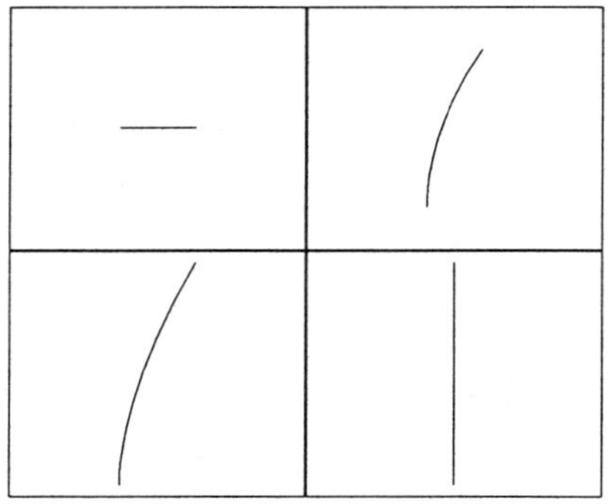

Figure 10.3.2.

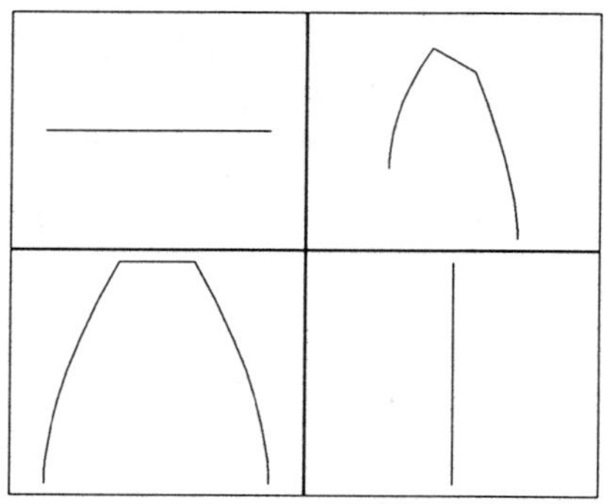

Figure 10.3.3.

```
'Program SSmithTP.BAS — Draws a gear tooth using involute by '
                  applying SilverScreen and BASIC.
   a1   = 95.0        : a2 = 60.0 : da = 3.5        : pi = 3.14159
 conv   = pi/180.0 : rb = 200.0 : a = a1            : t = a*conv
   xb   = rb*cos(t)    : yb = rb*sin(t)
FOR i  =  2.0 TO 11.0
     a  =  (a1-(i-1.0)*da)*conv
    rt  =  rb*(i-1.0)*da*conv
    xe  =  rb*cos(a)-rt*cos(a-pi/2.0)
    ye  =  rb*sin(a)-rt*sin(a-pi/2.0)
@draw line {xb},{yb},0.0 {xe},{ye},0.0
@draw line {-xb},{yb},0.0 {-xe},{ye},0.0
```

```
    xb = xe : yb = ye : NEXT i
  @draw line {xe},{yb},0.0 {-xe},{ye},0.0
  END
```

10.4 APPLYING SILVERSCREEN COMMANDS IN C PROGRAMS

SilverSmith also enables the **SilverScreen** commands to be applied in programs written in the **C** language. In this section, two **C** programs, **ConCir.C** and **RectGrid.C**, are presented to illustrate how concentric circles and rectangular grids can be plotted on screen by combining the **SilverScreen** and **C** statements.

PROGRAM ConCir.C

We have learned earlier how a circle can be drawn by using the options radius, diameter, circumference, tangent, and so on. Program **ConCir.C** listed below uses the commands **Draw**, **Circle**, and **Diameter** of **SilverScreen** to draw a series of six concentric circles having their centers all located at (0,0,0) and radii varying from 5 to 10, all in generic units. The command **for** in **C** language provides the needed looping of the radius *r*. The **@ at** statement within the loop moves the cursor to the specified location ({-*r*},0,0), which is the leftmost point of the circumference of the circle. After the circles have been drawn, the last **@** statement requests a four-way, split-window

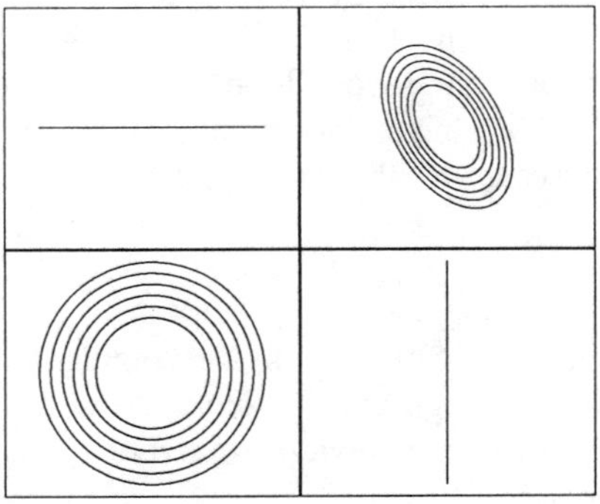

Figure 10.4.1.

display of the resulting plot shown in Figure 10.4.1.

As for the **BASIC** programs, **SilverSmith C** programs need to be compiled using the commands

Language Compile C filename (with extension **.C**)

An excutable file having the same file name and with an extension **.EX** is thus created. This executable file can then be run to generate the concentric circles continuously or by single steps on the screen by using the commands

Language Execute C Continuous/Single-step

The program **ConCir.C** is listed below.

```
main( )
{ /* Program ConCir.C — Draws concentric circles SilverScreen and C */
        double r,r1,r2 ; r1 = 5.0 ; r2 = 10.0 ;
        for (r = r1; r< = r2; ++r)
        {
                @ at {-r},0.0,0.0
                @ draw circle diameter {2*r},0.0,0.0
                }
        @ window split four-way drawing
}
```

PROGRAM RectGrid.C

As a second illustration of using the **SilverScreen** commands in a **C** program, a rectangular grid is generated by the C program called **RectGrid.C**. In this program, the commands **Draw** and **Line** of **SilverScreen** are applied in two loops using the **C** command **for**. The first loop draws 11 vertical lines for the *x* values ranging from 10 to 50 using an increment of 4; the second loop draws nine horizontal lines for the *y* values ranging from 30 to 70 using an increment of 5. Figure 10.4.2 shows the four-way, split-window display of the resulting rectangular grid. The program **RectGrid.C** is listed below.

```
main( )
{ /* Program RectGrid.C — SilverSmith Rectangular Grid in C
        language draws a rectangular grid from x1 to x2 and y1
        to y2 using increments xi and yi, respectively          */
double x,x1,x2,xi,y,y1,y2,yi;
        x1 = 10. ; x2 = 50. ; xi = 4. ; y1 = 30. ; y2 = 70. ; yi = 5. ;
for (x = x1; x < = x2;)
        { @ draw line {x},{y1},0.0 {x},{y2},0.0
        x = x + xi; }
for (y = y1; y < = y2;)
        { @ draw line {x1},{y},0.0 {x2},{y},0.0
        y = y + yi; }
@ window split four-way drawing
}
```

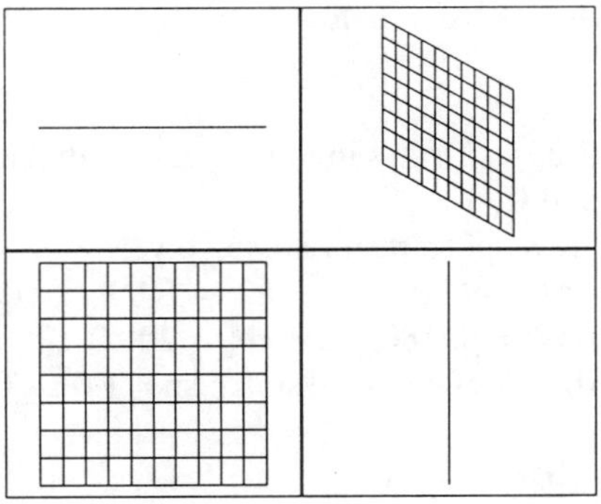

Figure 10.4.2.

10.5 MORE EXAMPLES OF SILVERSMITH-C PROGRAMS

In Section 10.3, the **BASIC** program **SSmithTP.BAS** draws an involute profile of a gear tooth. With very little change, we can easily develop it into a **C** program. Program **GTooth.C** is the result, which also includes an addition of a *base circle* of the gear. Also, instead of using two **Draw** and **Line** statements in the **BASIC** program **SSmithTP.BAS**, two nested loops (outer *j* loop and inner *i* loop) are employed in the **C** program **GTooth.C**. For $j = 1$ and $j = 2$, the latter program draws the left and right profiles, respectively, when the variable **sn** changes its sign from first positive and then negative. The horizontal line connecting the two profiles is drawn by the last **@** draw line statement. The **@ at** statement places the cursor at the center, (0,0,0), of the base circle and then the **Draw Circle Radius** statement draws the base circle using a radius equal to *rb*. The resulting four-way, split-window display of the one-tooth gear is shown in Figure 10.5.1.

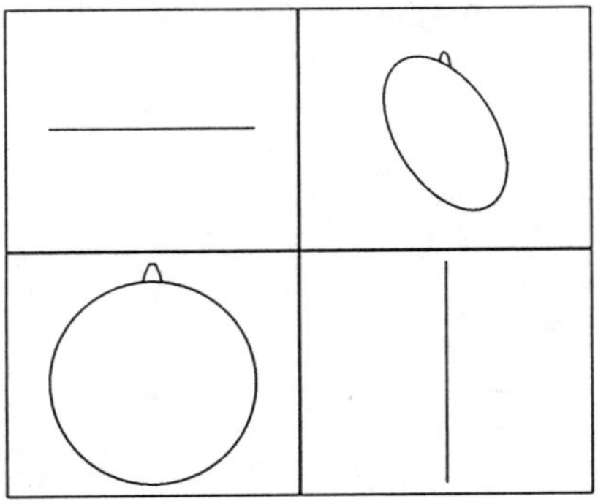

Figure 10.5.1.

The program **GTooth.C** is listed below.

```
main( )
{ /* Program GTooth.C — Draws involute gear-tooth with
     SilverScreen and C */
   double a,a1,a2,conv,da,pi,rb,rt,sn,t,xb,xe,yb,ye;
   int i,j          ;   a1  = 95.0     ;  a2  = 60.0  ;  da  = 3.5 ;
   pi = 3.14159 ; conv  = pi/180.0 ;  rb  = 200.0 ;   a  = a1 ;
t = a*conv ;   xb = rb*cos(t); yb = rb*sin(t); sn = 1.0 ;
 for (j = 1; j<= 2; ++j)
   { xb = rb*cos(t) ; yb = rb*sin(t) ;
              for (i = 2; i<= 11; ++i)
              { a = (a1-(i-1.0)*da)*conv ; rt = rb*(i-1)*da*conv ;
              xe = rb*cos(a)-rt*cos(a-pi/2.0);
              ye = rb*sin(a)-rt*sin(a-pi/2.0);
              @ draw line {sn*xb},{yb},0.0 {sn*xe},{ye},0.0
              xb = xe ; yb = ye ;          }
              sn = -sn;
              @ draw line {sn*xe},{ye},0.0 0.0,{ye},0.0
                                                     }
 @ at 0,0,0
 @ draw circle radius {rb},0,0
 @ window split four-way drawing
}
```

More can be added in the above program such as to rotate and copy the tooth and then place the copied tooth at a different angle on the base circle so that the result resembles a cross-sectional view of a gear as shown in Figure 10.5.2. Further enhancement can be made by application of the **Sweep** command in the **Geometry** menu of **SilverScreen** to form a three-dimensional solid model of a gear (Figure 10.5.3).

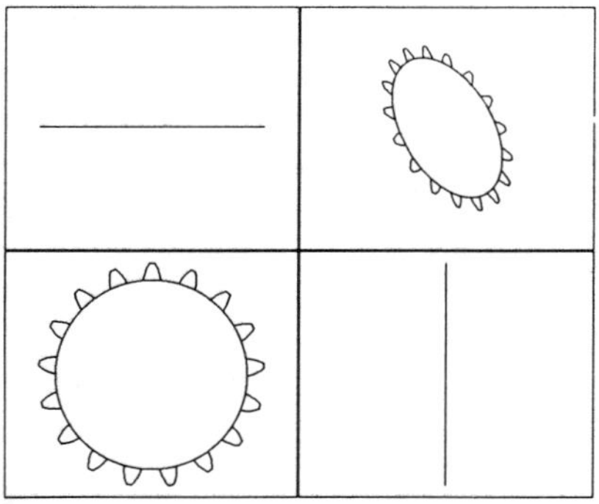

Figure 10.5.2.

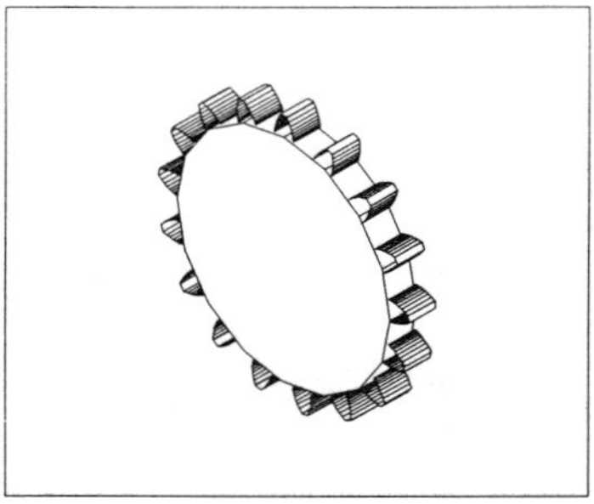

Figure 10.5.3.

PROBLEMS

10.1 Modify the program **SSmithC.BAS** by gradually reducing the angle increment so that the sides of the resulting polygon are increased and eventually approach the shape of a smooth circle.

10.2 Develop a **BASIC** program, in which a geometric profile of a cam[4] is defined (Figure P10.2 showing a series of different cams as commonly assembled on a cam shaft of an automobile). This cam profile is then displayed on screen using the **SilverScreen** commands.

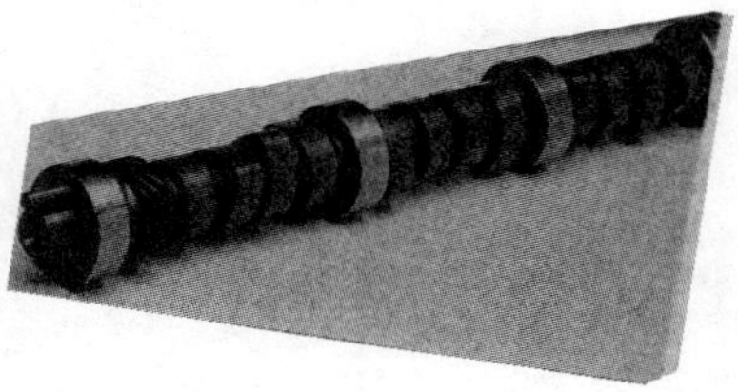

Figure P10.2.

10.3 Same as Problem 10.2 but using the **C** language.

10.4 Expand the program **GTooth.C** introduced in Section 10.5 to generate Figure 10.5.2.

10.5 Further expand the program **GTooth.C** so that Figure 10.5.3 can be displayed on screen.

10.6 Develop the **BASIC** version of Program **GTooth.C** so that both Figures 10.5.2 and 10.5.3 can be displayed using **SilverScreen**.

10.7 Create a data file containing 10 points, (X_i,Y_i) for $i = 1,2,...,10$, and then write a **BASIC** program that reads this data file and has appropriate **@ SilverSmith** statements to draw linear segements joining the adjacent points.

10.8 Same as Problem 10.9 but use the **C** language.

10.9 Expand the data file described in Problem 10.7 to have 5 sets of 10 points each for five different layers (in the third, Z dimension) $Z_j = j$ for $j = 1$ through 5. The curves constructed in two adjacent layers are to be connected in the **F-mesh** fashion.

10.10 Same as Problem 10.11 but use the **C** language.

10.11 Apply the hidden-line removal and shading capabilities of SilverScreen to display the three-dimensional surface generated by the program dveloped for Problem 10.11 or Problem 10.12.

10.12 Develop a BASIC or **C** program for generating an ellipsoid and then display it using the @ SilverSmith statements.

10.13 Use the FOR statement of the **C** language to generate the six sides of a hexagon and display it using the **@ SilverSmith** statements.

---

[4]See, for example, Y. C. Pao, *Elements of Computer-Aided Design and Manufacturing, CAD/CAM*, John Wiley, New York, 1984.

10.14 Use the **FOR** and **NEXT** statement of the **BASIC** language to implement Problem 10.13.

10.15 The clock shown in Figure P10.15 has numeric markings 1 through 12, which are best suited by applying the loop process in programming (such as the **FOR** and **NEXT** pair in the **BASIC** language, and **FOR** in the **C** language) method. Write a **BASIC** program to incorporate the **@** statments of **SilverSmith** for displaying the clock on screen.

Figure P10.15.

10.16 Same as Problem 10.15 but use the **C** language.

10.17 Develop a **BASIC** or **C** program using the looping programming concept to model the DNA strand shown in Figure P10.17.

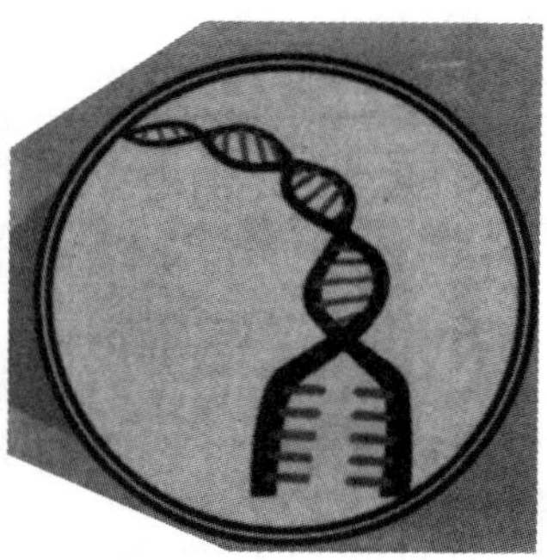

Figure P10.17.

APPENDIX A
SilverScreen Fill Patterns

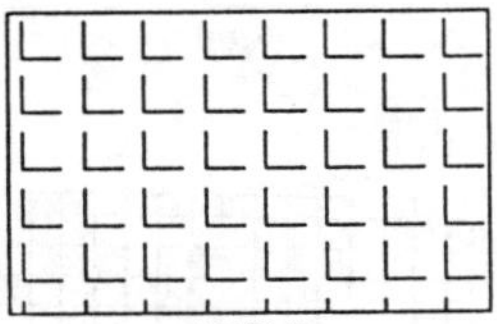
ANGLE

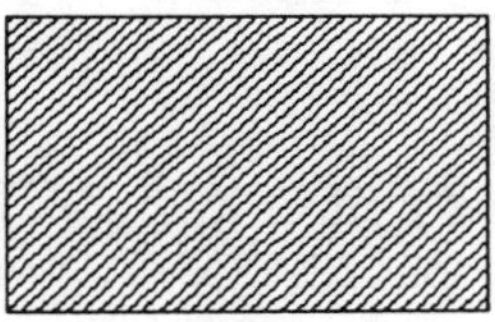
ANSI31

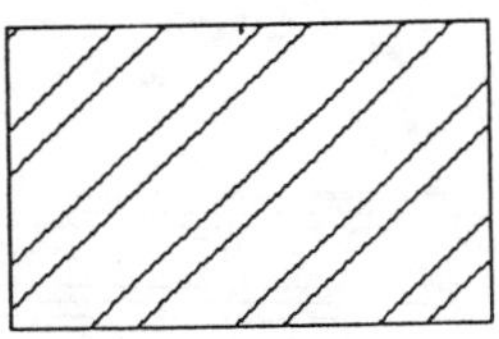
ANSI32

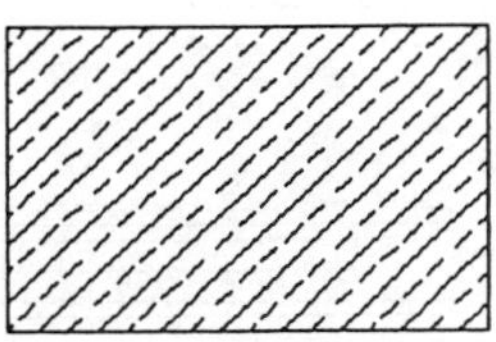
ANSI33

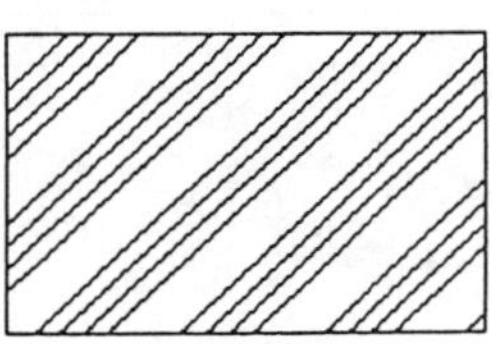
ANSI34

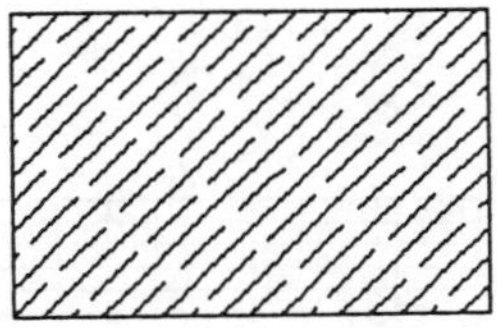
ANSI35

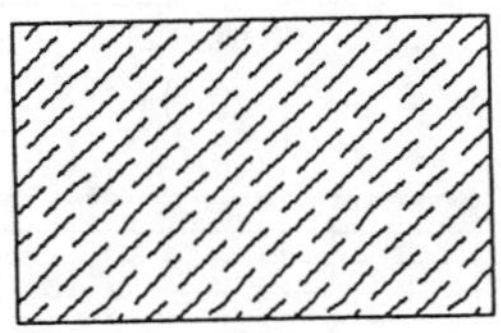
ANSI36

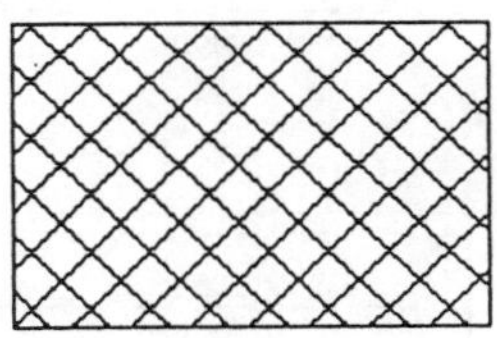
ANSI37

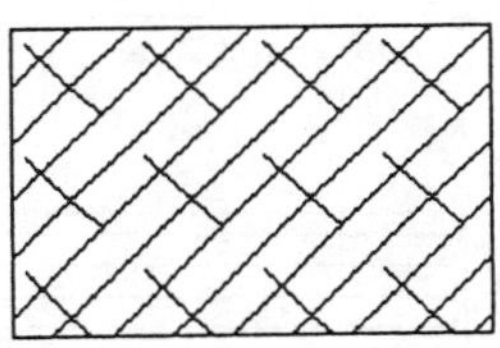
ANSI38

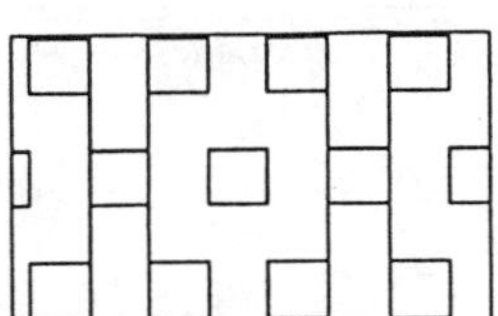
BOX

BRASS

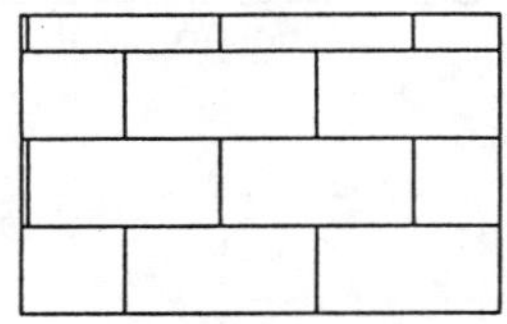
BRICK

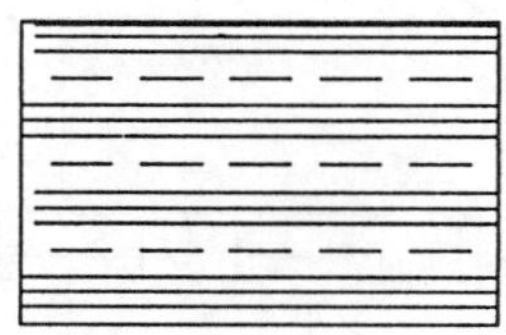
CLAY

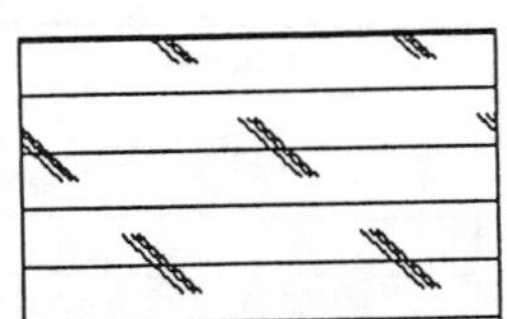
CORK

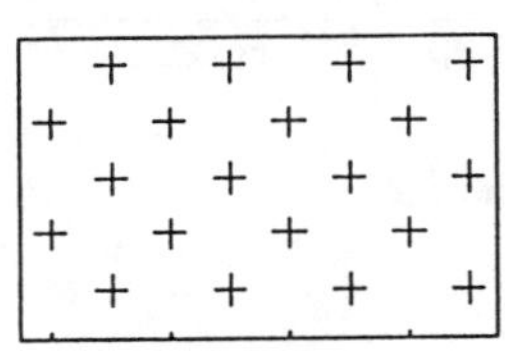
CROSS

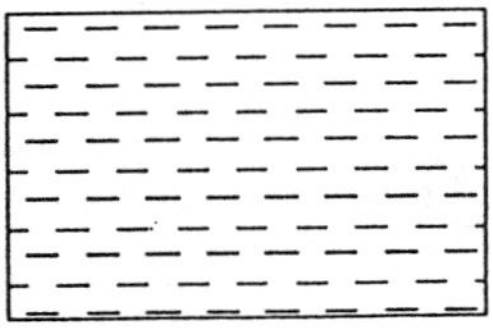
DASH

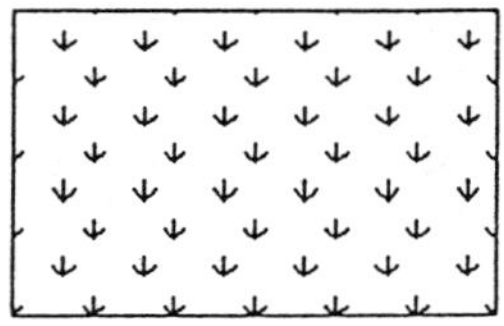
GRASS

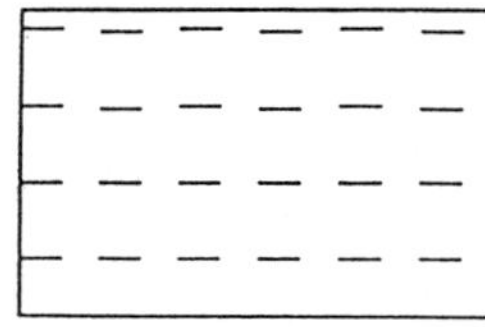
MUDST

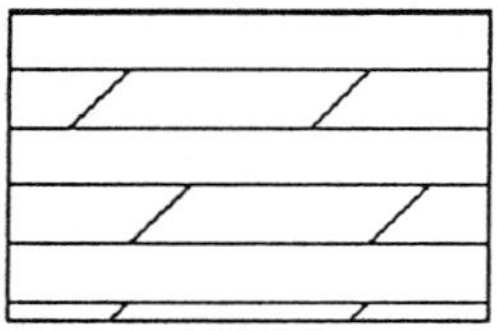
DOLMIT

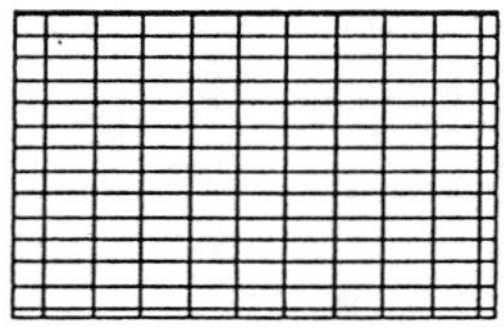
GRATE

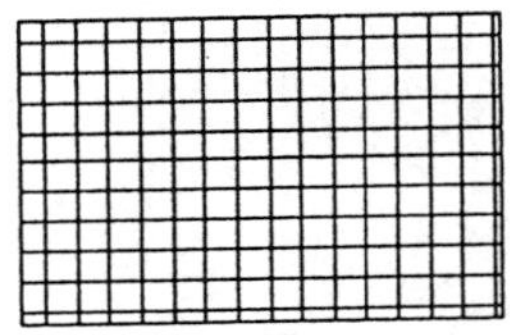
NET

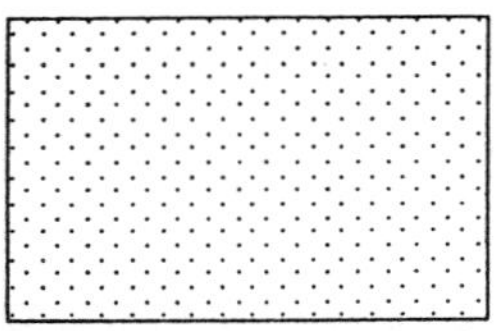
DOTS

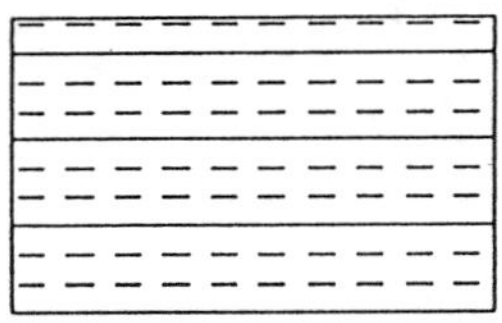
INSUL

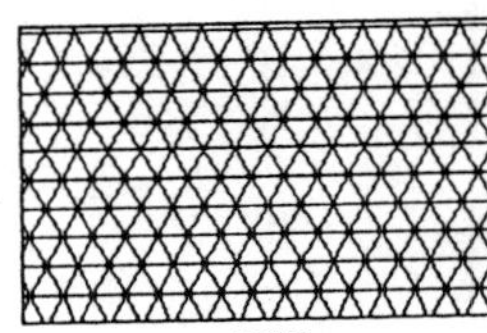
NET3

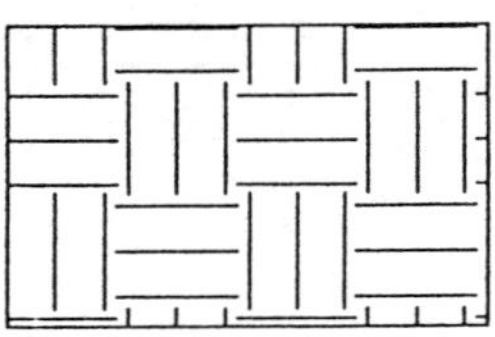
EARTH

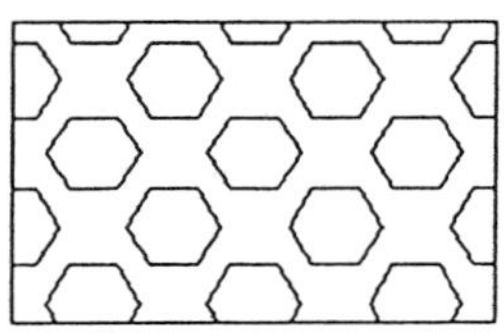
HEX

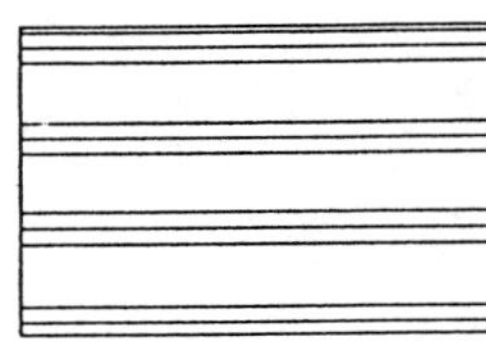
PLAST

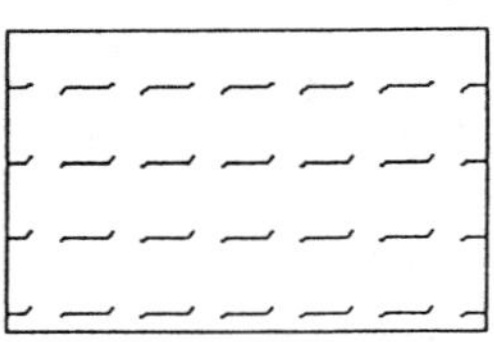
FLEX

LINE

PLASTI

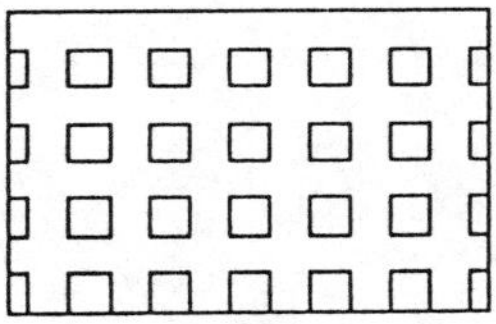

SQUARE

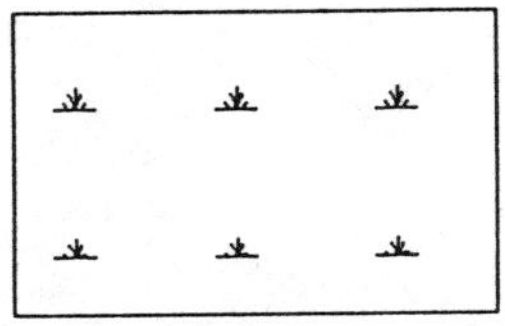

SWAMP

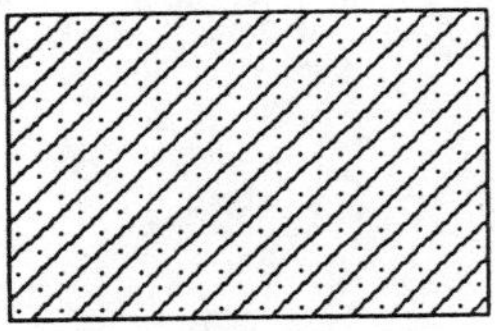

SACNCR

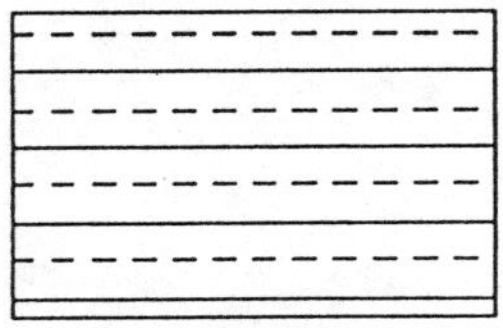

TRANS

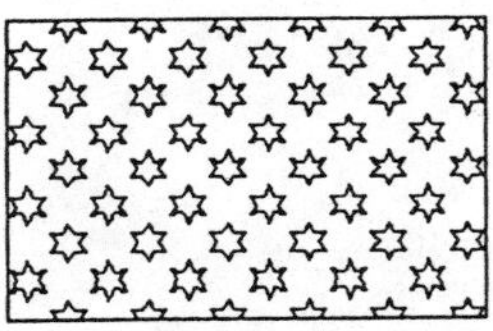

STARS

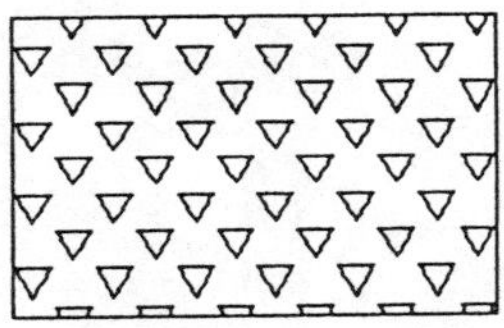

TRIANG

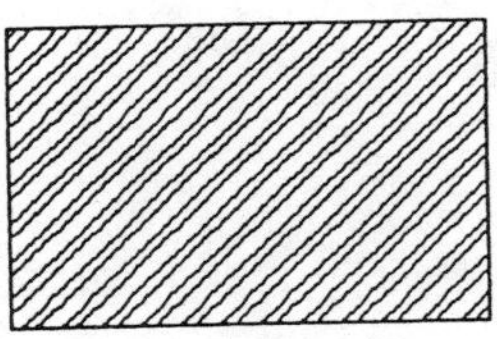

STEEL

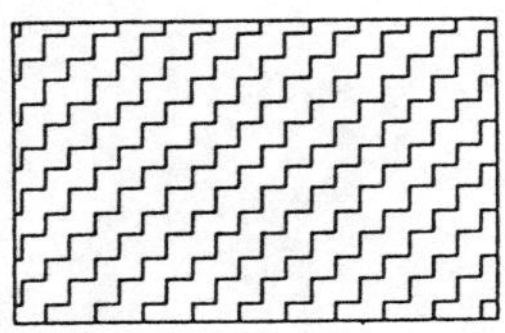

ZIGZAG

APPENDIX B
Sample Drafting Rules and Standards

A number of American Standard dimensioning rules and drafting symbols are adopted herein. Only for the purpose of illustration, some tables may not be complete. Other drafting textbooks should be consulted if the reader cannot find a particular table.

American Standard (ANSI B17.1) Plain and Gib-Head Keys

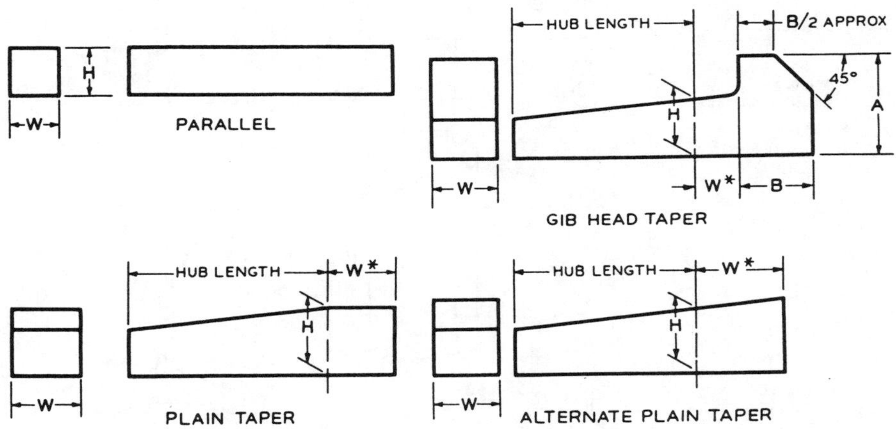

Plain and Gib Head Taper Keys Have a 1/8" Taper in 12"

KEY			NOMINAL KEY SIZE Width, W		TOLERANCE	
			Over	To (Incl)	Width, W	Height, H
Parallel	Square	Bar Stock	–	3/4	+0.000 –0.002	+0.000 –0.002
			3/4	1-1/2	+0.000 –0.003	+0.000 –0.003
			1-1/2	2-1/2	+0.000 –0.004	+0.000 –0.004
			2-1/2	3-1/2	+0.000 –0.006	+0.000 –0.006
		Keystock	–	1-1/4	+0.001 –0.000	+0.001 –0.000
			1-1/4	3	+0.002 –0.000	+0.002 –0.000
			3	3-1/2	+0.003 –0.000	+0.003 –0.000
	Rectangular	Bar Stock	–	3/4	+0.000 –0.003	+0.000 –0.003
			3/4	1-1/2	+0.000 –0.004	+0.000 –0.004
			1-1/2	3	+0.000 –0.005	+0.000 –0.005
			3	4	+0.000 –0.006	+0.000 –0.006
			4	6	+0.000 –0.008	+0.000 –0.008
			6	7	+0.000 –0.013	+0.000 –0.013
		Keystock	–	1-1/4	+0.001 –0.000	+0.005 –0.005
			1-1/4	3	+0.002 –0.000	+0.005 –0.005
			3	7	+0.003 –0.000	+0.005 –0.005
Taper	Plain or Gib Head Square or Rectangular		–	1-1/4	+0.001 –0.000	+0.005 –0.000
			1-1/4	3	+0.002 –0.000	+0.005 –0.000
			3	7	+0.003 –0.000	+0.005 –0.000

*For locating position of dimension *H*. Tolerance does not apply.
All dimensions given in inches.

American Standard (ANSI Z32.2.3) Graphical Symbols for Pipe Fittings and Valves

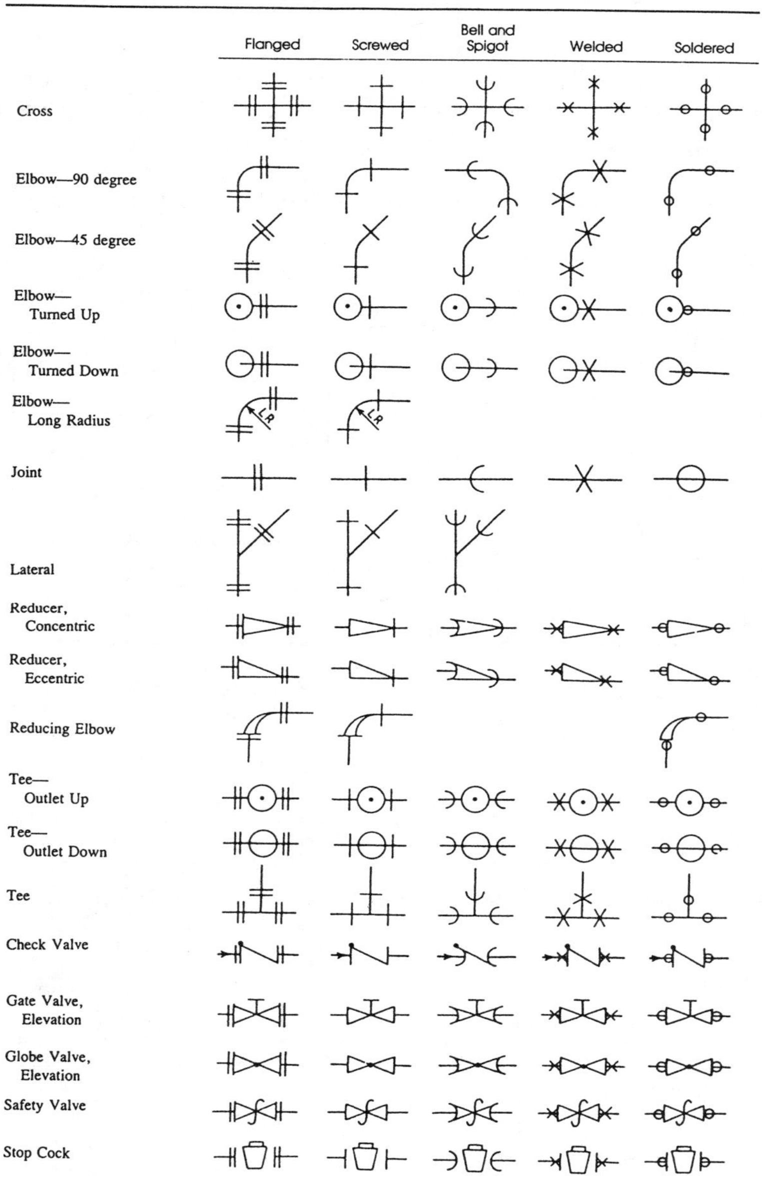

Standard Symbols for the Most Common Electrical and Electronic Components

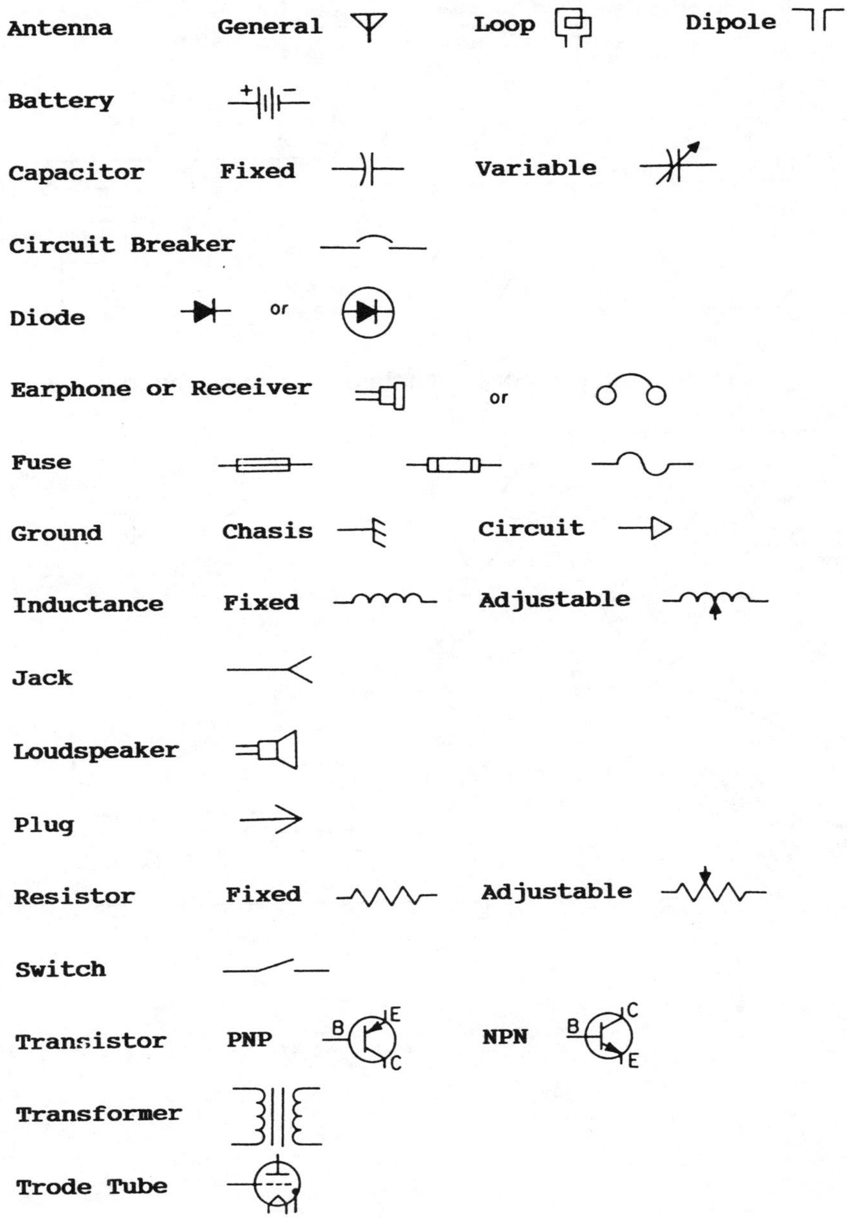

American Standard Weld Symbols

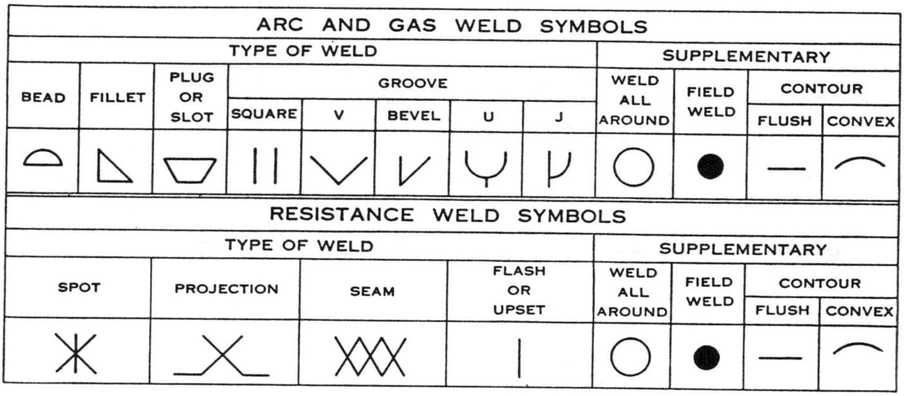

Standard Location of Elements of a Welding Symbol

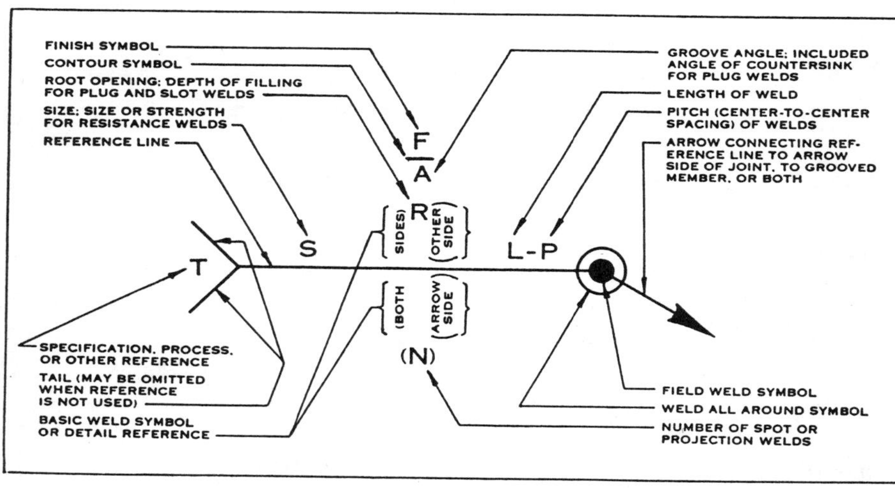

American National Standards Institute B.18.2.1 ASTM* and SAE** Grade Markings for Steel Bolts and Screws

Grade Marking	Specification	Material
 NO MARK	SAE–Grade 1	Low or Medium Carbon Steel
	ASTM–A307	Low Carbon Steel
	SAE–Grade 2	Low or Medium Carbon Steel
	SAE–Grade 5	Medium Carbon Steel, Quenched and Tempered
	ASTM–A 449	
	SAE–Grade 5.2	Low Carbon Martensite Steel, Quenched and Tempered
	ASTM–A 325 Type 1	Medium Carbon Steel, Quenched and Tempered Radial dashes optional
	ASTM–A 325 Type 2	Low Carbon Martensite Steel, Quenched and Tempered
	ASTM–A 325 Type 3	Atmospheric Corrosion (Weathering) Steel, Quenched and Tempered
	ASTM–A 354 Grade BC	Alloy Steel, Quenched and Tempered
	SAE–Grade 7	Medium Carbon Alloy Steel, Quenched and Tempered, Roll Threaded After Heat Treatment
	SAE–Grade 8	Medium Carbon Alloy Steel, Quenched and Tempered
	ASTM–A 354 Grade BD	Alloy Steel, Quenched and Tempered
	SAE–Grade 8.2	Low Carbon Martensite Steel, Quenched and Tempered
	ASTM–A 490 Type 1	Alloy Steel, Quenched and Tempered
	ASTM–A 490 Type 3	Atmospheric Corrosion (Weathering) Steel, Quenched and Tempered

* American Society for Testing Materials.
** Society of Automotive Engineers.

American National Standard Hex Bolts

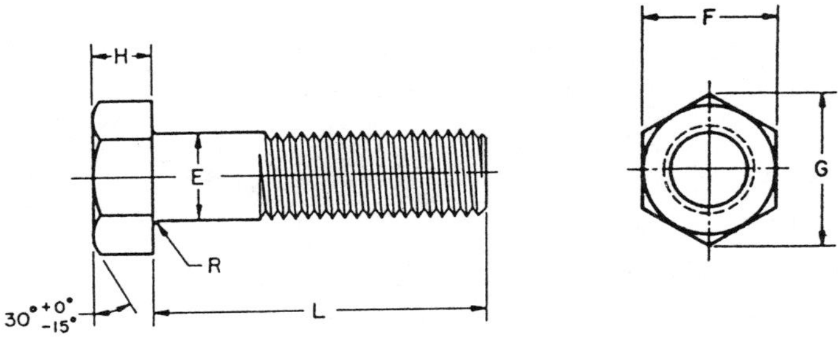

IFI506 – Dimensions of Metric Hex Bolts

Bolt Size & Thread Pitch	E		F		G		H		R
	Max	Min	Max	Min	Max	Min	Max	Min	Min
M5×0.8	5.48	4.52	8.00	7.75	9.24	8.84	3.88	3.35	0.2
M6.3×1	6.78	5.72	10.00	9.69	11.55	11.05	4.70	4.13	0.3
M8×1.25	8.58	7.42	13.00	12.60	15.01	14.36	5.73	5.10	0.4
M10×1.5	10.58	9.42	15.00	14.50	17.32	16.53	6.86	6.17	0.4
M12×1.75	12.70	11.30	18.00	17.40	20.78	19.84	7.99	7.24	0.4
M14×2	14.70	13.30	21.00	20.30	24.25	23.14	9.32	8.51	0.6
M16×2	16.70	15.30	24.00	23.20	27.71	26.45	10.56	9.68	0.6
M20×2.5	20.84	19.16	30.00	29.00	34.64	33.06	13.12	12.12	0.8
M24×3	24.84	23.16	36.00	34.80	41.57	39.67	15.68	14.56	0.8
M30×3.5	30.84	29.16	46.00	44.50	53.12	50.73	19.48	17.92	1.2
M36×4	37.00	35.00	55.00	53.20	63.51	60.65	23.38	21.72	1.2
M42×4.5	43.00	41.00	65.00	62.90	75.06	71.71	26.97	25.03	1.2
M48×5	49.00	47.00	75.00	72.60	86.60	82.76	31.07	28.93	1.5
M56×5.5	57.20	54.80	85.00	82.20	98.15	93.71	36.20	33.80	2.0

APPENDIX C

SilverScreen Menus, Commands, and Quickkey Assignments

QUICKKEY ASSIGNMENTS

Function Keys

F1 Accesses to the context sensitive to on-line help. The help displayed is related to the currently highlighted menu option.

F2 Displays the currently loaded custom menu.

F3 Refreshes the drawing in all windows of the current screen.

F4 Makes the nest window current when more than one window is displayed.

F5 Toggles display of the current window between parallel and perspective projection.

F6 Changes to the next screen if more than one screen is active.

F7 Prompts for the name of a macro to be executed.

F8 Prompts for the name of a program or script file to be executed.

F9 Repeats the previous command.

F10 Ends the current drawing session.

Alternate Keys

Alt A Toggles absolute and relative modes of coordinate entry.

Alt B Positions the lower-left corner of the construction space and grid at the current cursor location. The grid maintains its current orientation.

Alt C Changes display of the cursor to box, cross-hair, small cross, or triad.

Alt G Enables or disables the grid.

Alt I Prompts for entry of a new snap interval value.

Alt N Prompts for input of a new object name.

Alt Q Displays the Quick Menu in the menu prompt area.
Alt S Enables or disables the snap.

Alt T Defines a temporary object. Each temporary object is named in sequence and the current name is displayed in the status area.

Alt V Aligns the construction space to the view in the current window. The construction space *xy*-plane is set parallel to the view plane.

Alt W Toggles status area display and coordinate entry between construction and world spaces. The status area *X*:, *Y*:, and *Z*: indicators are highlighted.

Alt X Prompts for entry of a construction space *x* coordinate value. The absolute/relative mode setting determines whether the *x* value moves the cursor to an absolute or relative position.

Alt Y Prompts for entry of a construction space *y* coordinate value. The absolute/relative mode setting determines whether the *y* value moves the cursor to an absolute or relative position.

Alt Z Prompts for entry of a construction space *z* coordinate value. The absolute/relative mode setting determines whether the *z* value moves the cursor to an absolute or relative position.

Control Keys

CTRL C Places the cursor at the center of the current view.

CTRL -> The control right arrow keys select the next object. The object is highlighted and the name appears in the status area.

CTRL <- The control left arrow keys select the previous object. The object is highlighted and the name appears in the status area.

CTRL HM The control home keys select the next block. The block is highlighted and the name appears in the status area.

CTRL END The control end keys select the previous block. The block is highlighted and the name appears in the status area.

Contents